edexcel
advancing learning, changing lives

T0173447

Mechanics 4

# Edexcel AS and A-level Modular Mathematics

Susan Hooker
Mick Jennings
Jean Littlewood
Bronwen Moran
Laurence Pateman

# Contents

# About this book

This book is designed to provide you with the best preparation possible for your Edexcel M4 unit examination:

- The LiveText CD-ROM in the back of the book contains even more resources to support you through the unit.

Brief chapter overview and 'links' to underline the importance of mathematics: to the real world, to your study of further units and to your career

## Finding your way around the book

Detailed contents list shows which parts of the M4 specification are covered in each section

Every few chapters, a review exercise helps you consolidate your learning

After completing this chapter you should be able to:
- investigate the motion of a particle which is moving under the influence of a restoring force proportional to the particle's displacement and a resistance which is proportional to its speed
- investigate the motion of a particle which is moving under the influence of the above two forces and is also forced to oscillate with a frequency other than its natural one.

# 4

# Damped and forced harmonic motion

You studied simple harmonic motion (S.H.M.) in book M3, Chapter 3. When a particle is attached to an elastic string or spring and set in motion with no forces other than the tension/thrust and gravity acting on it, the motion is simple harmonic. The simple harmonic oscillations have constant amplitude. However, in practice, the amplitude of the oscillations decreases and the particle comes to rest fairly quickly. Hence the model must be refined to incorporate other factors such as air resistance and friction.

The toy in the picture can be held down and released. It will oscillate about its equilibrium position but the oscillations will not continue indefinitely. Soon after release the 'man' will come to rest.

## Contents

Each section begins with a statement of what is covered in the section

Concise learning points

Step-by-step worked examples

### 3.1 You can use calculus when a particle moves in a straight line against a resistance which varies with the speed of the particle.

The displacement from a fixed point (x), velocity (v) and acceleration (a) when the acceleration of a particle is varying with time are connected by the relations

$$a = \frac{dv}{dt} = \frac{d^2x}{dt^2} \qquad ①$$

Using the chain rule for differentiation

$$a = \frac{dv}{dt} = \frac{dv}{dx} \times \frac{dx}{dt}$$

As $v = \frac{dx}{dt}$, then

$$a = \frac{dv}{dx} \times v = v\frac{dv}{dx} \qquad ②$$

Also, if you differentiate $\frac{1}{2}v^2$ implicitly with respect to $x$, you obtain

$$\frac{d}{dx}(\frac{1}{2}v^2) = \frac{1}{2} \times 2v \times \frac{dv}{dx} = v\frac{dv}{dx} \qquad ③$$

Combining the results ② and ③, you obtain $a = v\frac{dv}{dx} = \frac{d}{dx}(\frac{1}{2}v^2)$.

The alternative forms for the acceleration can be summarised

■ $a = \frac{dv}{dt} = \frac{d^2x}{dt^2} = v\frac{dv}{dx} = \frac{d}{dx}(\frac{1}{2}v^2)$

With these alternative forms for the acceleration you can solve problems where the resistance to the motion of a particle varies with its speed or velocity. In this chapter, you will mainly use the forms $\frac{dv}{dt}$ and $v\frac{dv}{dx}$.

If the resistance is a function of velocity f(v), then the equation of motion of the particle will often be a separable differential equation which you can solve using the method you learnt in book C4. In this book, the function f(v) will always be in the form $a + bv$ or $a + bv^2$, where $a$ and $b$ are constants.

Displacements must be measured from a fixed point. You will often choose to measure displacements from the point from which a particle starts or the point from which it is projected.

■ In forming an equation of motion, forces that tend to decrease the displacement are negative and forces that tend to increase the displacement are positive.

### Example 1

A particle P of mass 0.5 kg moves in a straight horizontal line. When the speed is $v$ m s⁻¹ the resultant force acting on P is a resistance of magnitude 3v N.
Find the distance moved by P as it slows down from 12 m s⁻¹ to 6 m s⁻¹.

You measure the displacement of P, x m, from the point where it has a speed of 12 m s⁻¹.

$R(\rightarrow) F = ma$

$-3v = 0.5a$

As the resistance to motion is acting in the direction which decreases the displacement x m, the term, 3v, representing the resistance in the equation of motion, has a negative sign.

$-3v = 0.5v\frac{dv}{dx}$

$\frac{dv}{dx} = -6$

When the question asks you to relate distance with speed, you choose the expression $a = v\frac{dv}{dx}$ for the acceleration.

$v = \int -6 \, dx$

$v = -6x + A$

At $x = 0, v = 12$

$12 = -6 \times 0 + A \Rightarrow A = 12$

Hence

$v = 12 - 6x$

The displacement is measured from the point where the speed of P is 12 m s⁻¹. So you evaluate the constant of integration using x = 0 when v = 12.

When $v = 6$

$6 = 12 - 6x$

$x = \frac{12-6}{6} = 1$

The distance moved by P as it slows down from 12 m s⁻¹ to 6 m s⁻¹ is 1 m.

### Exercise 3A

1 A particle P of mass 2.5 kg moves in a straight horizontal line. When the speed of P is $v$ m s⁻¹, the resultant force acting on P is a resistance of magnitude 10v N. Find the time P takes to slow down from 24 m s⁻¹ to 6 m s⁻¹.

2 A particle P of mass 0.8 kg is moving along the axis Ox in the direction of x-increasing. When the speed of P is $v$ m s⁻¹, the resultant force acting on P is a resistance of magnitude $0.4v^2$ N. Initially P is at O and is moving with speed 12 m s⁻¹. Find the distance P moves before its speed is halved.

# 1

# Review Exercise

1 A river of width 40 m flows with uniform and constant speed between straight banks. A swimmer crosses as quickly as possible and takes 30 s to reach the other side. She is carried 25 m downstream. Find
  a the speed of the river,
  b the speed of the swimmer relative to the water.

2 At noon, a boat P is on a bearing of 120° from boat Q. Boat P is moving due east at a constant speed of 12 km h⁻¹. Boat Q is moving in a straight line with a constant speed of 15 km h⁻¹ on a course to intercept P. Find the direction of motion of Q, giving your answer as a bearing.

3 Points A and B are directly opposite each other on the parallel banks of a river. A motorboat, which travels at 4 m s⁻¹ relative to the water, crosses from A to B. Given that the distance AB is 400 m and that the river is flowing at 1.5 m s⁻¹ parallel to the banks, calculate

  a the angle, to the nearest degree, between AB and the direction in which the boat is being steered,
  b the speed, in m s⁻¹ to 2 significant figures, of the motorboat relative to the bank,
  c the time, to the nearest second, taken by the motorboat to cross the river.

4 A boy enters a large horizontal field and sees a friend 100 m due north. The friend is walking in an easterly direction at a constant speed of 0.75 m s⁻¹. The boy can walk at a maximum speed of 1 m s⁻¹.
  Find the shortest time for the boy to intercept his friend and the bearing on which he must travel to achieve this.

5 A cyclist P is cycling due north at a constant speed of 20 km h⁻¹. At 12 noon another cyclist Q is due west of P. The speed of Q is constant at 10 km h⁻¹.
  Find the course which Q should set in order to pass as close to P as possible, giving your answer as a bearing.

Past examination questions are marked 'E'

Each section ends with an exercise – the questions are carefully graded so they increase in difficulty and gradually bring you up to standard

Each chapter has a different colour scheme, to help you find the right chapter quickly

Each chapter ends with a mixed exercise and a summary of key points.

At the end of the book there is an examination-style paper.

## LiveText software

The LiveText software gives you additional resources: Solutionbank and Exam café. Simply turn the pages of the electronic book to the page you need, and explore!

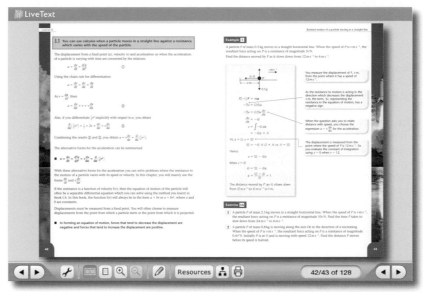

## Unique Exam café feature:

- Relax and prepare – revision planner; hints and tips; common mistakes
- Refresh your memory – revision checklist; language of the examination; glossary
- Get the result! – fully worked examination-style paper

## Solutionbank

- Hints and solutions to every question in the textbook
- Solutions and commentary for all review exercises and the practice examination paper

Published by Pearson Education Limited, a company incorporated in England and Wales, having its registered office at 80 strand, London, WC2R 0RL. Registered company number: 872828
www.pearsonschoolsandfecolleges.co.uk

Edexcel is a registered trademark of Edexcel Limited

Text © Susan Hooker, Mick Jennings, Jean Littlewood, Bronwen Moran, Laurence Pateman 2009

15
10 9 8 7

British Library Cataloguing in Publication Data is available from the British Library on request.

ISBN 978 0 435519 24 7

Edited by Susan Gardner
Typeset by Tech-Set Ltd, Gateshead
Illustrated by Tech-Set Ltd, Gateshead
Cover design by Christopher Howson
Picture research by Chrissie Martin
Cover photo/illustration © Science Photo Library/Laguna Design
Index by Indexing Specialists (UK) Ltd
Printed in China (SWTC/07)

**Acknowledgements**
The author and publisher would like to thank the following individuals and organisations for permission to reproduce photographs:

Shutterstock/HFNG **p**1; Press Association Images/Rui Vieira/PA Wire **p**24; Getty Images/PhotoDisc **p**41; Alamy Images/Tompiodesign **p**73; Alamy Images/ICP **p91r**; Alamy Images/Nigel Lloyd **p91l**.

Every effort has been made to contact copyright holders of material reproduced in this book. Any omissions will be rectified in subsequent printings if notice is given to the publishers.

**Disclaimer**
This Edexcel publication offers high-quality support for the delivery of Edexcel qualifications.

Edexcel endorsement does not mean that this material is essential to achieve any Edexcel qualification, nor does it mean that this is the only suitable material available to support any Edexcel qualification. No endorsed material will be used verbatim in setting any Edexcel examination/assessment and any resource lists produced by Edexcel shall include this and other appropriate texts.

Copies of official specifications for all Edexcel qualifications may be found on the Edexcel website - www.edexcel.com

After completing this chapter you should be able to:

- solve problems involving interception and collision
- solve problems involving the closest approach of two moving bodies.

In this chapter you will find out how to deal with the motion of bodies which are moving relative to each other.

# Relative motion

These two express trains are both moving at high speed but, relative to each other, they are moving quite slowly.

## 1.1 You can use and understand relative displacement.

■ The **position vector, $r_p$, of a point $P$ relative to a fixed origin $O$** was defined, in earlier modules, to be given by $\mathbf{r}_p = \overrightarrow{OP}$.

■ We can extend this idea by defining the **position vector of a point $P$ relative to another point $Q$**, written $_p\mathbf{r}_Q$, as being given by

$$_p\mathbf{r}_Q = \overrightarrow{QP}$$

(This is sometimes referred to as **the relative displacement of $P$ from $Q$**.)

The diagram shows that, if $O$ is the origin,

then

$$_p\mathbf{r}_Q = \overrightarrow{QP} = -\mathbf{r}_Q + \mathbf{r}_p = \mathbf{r}_p - \mathbf{r}_Q$$

i.e.

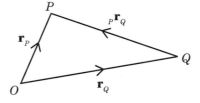

■ $_p\mathbf{r}_Q = \mathbf{r}_p - \mathbf{r}_Q$

The distance $PQ = |_p\mathbf{r}_Q| = |_Q\mathbf{r}_P|$.

### Example 1

If $A$ is the point $(4, -2)$ and $B$ is the point $(-1, 3)$, find

**a** the position vector of $A$ relative to $B$,

**b** the position vector of $B$ relative to $A$,

**c** the distance $AB$.

**a** $_A\mathbf{r}_B = \mathbf{r}_A - \mathbf{r}_B$

$$= \begin{pmatrix} 4 \\ -2 \end{pmatrix} - \begin{pmatrix} -1 \\ 3 \end{pmatrix}$$

$$= \begin{pmatrix} 5 \\ -5 \end{pmatrix}$$

> Note the order of the letters $A$ and $B$ is the same on both sides. $B$ is the origin here.

**b** $_B\mathbf{r}_A = \mathbf{r}_B - \mathbf{r}_A$

$$= \begin{pmatrix} -1 \\ 3 \end{pmatrix} - \begin{pmatrix} 4 \\ -2 \end{pmatrix}$$

$$= \begin{pmatrix} -5 \\ 5 \end{pmatrix}$$

> $_B\mathbf{r}_A = -_A\mathbf{r}_B$ by definition.

**c** $AB = |_A\mathbf{r}_B| = |_B\mathbf{r}_A|$

$$= \sqrt{(-5)^2 + 5^2}$$

$$= 5\sqrt{2}$$

## 1.2 You can solve problems that involve relative velocity.

From Section 1.1,

$$_P\mathbf{r}_Q = \mathbf{r}_P - \mathbf{r}_Q.$$

If we differentiate both sides of this vector equation with respect to time:

$$\frac{d}{dt}(_P\mathbf{r}_Q) = \frac{d}{dt}\mathbf{r}_P - \frac{d}{dt}\mathbf{r}_Q$$

■ $_P\mathbf{v}_Q = \mathbf{v}_P - \mathbf{v}_Q$ ← This is vector subtraction.

where $\mathbf{v}_P$ is the **velocity** of $P$, $\mathbf{v}_Q$ is the **velocity** of $Q$ and $_P\mathbf{v}_Q$ is the **velocity of $P$ relative to $Q$**.

■ $_P\mathbf{v}_Q$ shows **how $P$ would appear to be moving to an observer on $Q$**.

### Example 2

A ship $A$ is sailing due north at $20\,\text{km h}^{-1}$. Another ship $B$ is sailing due east at $15\,\text{km h}^{-1}$.

**a** Find the velocity of $A$ relative to $B$.

**b** Find the relative speed of $A$ to $B$.

**c** Find the direction that $B$ would appear to be moving in to an observer on ship $A$.

**a** We require $_A\mathbf{v}_B$, where

$$_A\mathbf{v}_B = \mathbf{v}_A - \mathbf{v}_B$$

$$|_A\mathbf{v}_B| = \sqrt{15^2 + 20^2}$$

$$= 25$$

$$\tan\theta = \frac{15}{20} = \frac{3}{4}$$

$$\theta = 36.9° \text{ (3 s.f.)}$$

We need to calculate the **magnitude** and **direction** of $_A\mathbf{v}_B$.

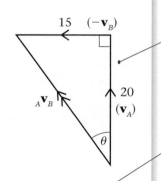

This is vector subtraction and we have no components, so we need to draw a vector triangle to perform this subtraction.

This means that, to an observer on ship $B$, ship $A$ would appear to be moving with speed $25\,\text{km h}^{-1}$ on a bearing of $323°$.

The relative speed is the magnitude of the relative velocity.

Thus the velocity of $A$ relative to $B$ is $25\,\text{km h}^{-1}$ on a bearing of $323°$.

**b** From above, the relative speed of $A$ to $B$ is $25\,\text{km h}^{-1}$.

**c** Since $_B\mathbf{v}_A = -\,_A\mathbf{v}_B$, the direction that $B$ would appear to be moving in to an observer on ship $A$ is on a bearing $143°$.

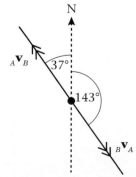

The velocity of $B$ relative to $A$ is in the opposite direction to that of $A$ relative to $B$.

## Example 3

The velocity of $P$ relative to $Q$ is $(3\mathbf{i} - \mathbf{j})\,\mathrm{m\,s^{-1}}$ and the velocity of $Q$ relative to $R$ is $(4\mathbf{i} + 5\mathbf{j})\,\mathrm{m\,s^{-1}}$. Find the velocity of $R$ relative to $P$.

$${}_P\mathbf{v}_Q = \mathbf{v}_P - \mathbf{v}_Q = (3\mathbf{i} - \mathbf{j})$$

$${}_Q\mathbf{v}_R = \mathbf{v}_Q - \mathbf{v}_R = (4\mathbf{i} + 5\mathbf{j})$$

Adding these two equations,

$$\mathbf{v}_P - \mathbf{v}_R = 7\mathbf{i} + 4\mathbf{j}$$

i.e. $\quad {}_P\mathbf{v}_R = 7\mathbf{i} + 4\mathbf{j}$

Hence, ${}_R\mathbf{v}_P = (-7\mathbf{i} - 4\mathbf{j})\,\mathrm{m\,s^{-1}}$

We could have used column vectors instead.

Eliminate $\mathbf{v}_Q$.

There is no need for a vector triangle as we can simply add the components.

Use ${}_R\mathbf{v}_P = -\,{}_P\mathbf{v}_R$.

## 1.3 You can solve problems which involve water current and wind effects.

If a swimmer is attempting to swim across a river in which there is a current flowing, or an aeroplane is flying and there is a wind blowing, we can use the ideas of relative velocity to determine the motion in each case.

## Example 4

A man who can row at $2.5\,\mathrm{m\,s^{-1}}$ in still water wishes to cross to the nearest point on the opposite bank of a river which is 200 m wide.

If the river is running at $1.5\,\mathrm{m\,s^{-1}}$,

**a** in which direction should he row,

**b** how many minutes does it take him to cross?

**c** If he wishes to cross as quickly as possible, in which direction should he row and where will he land?

There are 3 velocities involved:

$\mathbf{v}_M$ velocity of the man (relative to the Earth)

$\mathbf{v}_W$ velocity of the water (relative to the Earth)

${}_M\mathbf{v}_W$ velocity of the man (relative to the water)

These velocities will all each have a **magnitude** and **direction**.

It is essential that we set up some sensible notation.

This is the velocity of the man in **still water**.

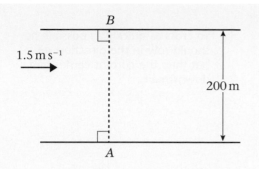

| | Magnitude | Direction |
|---|---|---|
| $\mathbf{v}_M$ | ? | Along AB |
| $\mathbf{v}_W$ | 1.5 | Parallel to bank |
| $_M\mathbf{v}_W$ | 2.5 | ? |

A simple diagram is always a good idea.

A is the initial position of the man.

It is always a good idea to use this type of table to **summarise** all the information given in the question.

Now,

$$_M\mathbf{v}_W = \mathbf{v}_M - \mathbf{v}_W$$

i.e.   $\mathbf{v}_M = \mathbf{v}_W + {}_M\mathbf{v}_W$

We now draw a vector triangle:

Remember this is vector subtraction.

Since we know the **direction** of $\mathbf{v}_M$ it is sensible to make this the subject.

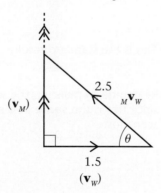

Draw $\mathbf{v}_W$ **first**, as we know **both** its magnitude and direction.

**a**   The direction in which the man rows is given by the direction of $_M\mathbf{v}_W$

$$\cos\theta = \frac{1.5}{2.5} = \frac{3}{5}$$

$$\theta = 53.1°$$

He rows upstream at 53.1° to the bank.

The man steers upstream so that the current then combines with his velocity and he actually proceeds straight across the river.

**b**   $|\mathbf{v}_M| = \sqrt{2.5^2 - 1.5^2}$

$$= 2\,\mathrm{m\,s^{-1}}$$

Time to cross $= \dfrac{200}{2}$

$$= 100\,\mathrm{s}$$

$$= 1\tfrac{2}{3}\ \text{minutes}$$

Use Pythagoras on the vector triangle.

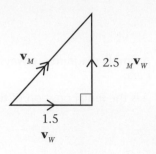

$$\text{Time to cross} = \frac{200}{2.5}$$
$$= 80 \, s$$
$$\text{Distance carried downstream} = 1.5 \times 80$$
$$= 120 \, m$$

To cross as quickly as possible he should row in the direction *AB* but then the current carries him downstream.

The velocity of the man (relative to the Earth), $\mathbf{v}_M$, has two components, $2.5 \, \text{m s}^{-1}$ across the river, and $1.5 \, \text{m s}^{-1}$ down the river.

## Example **5**

To a man cycling due E the wind appears to come from a direction N10°W; when he travels due N at the same speed as before the wind appears to be blowing from the direction N30°W. Find the actual direction of the wind.

$${}_W\mathbf{v}_M = \mathbf{v}_W - \mathbf{v}_M$$
$$\Rightarrow \quad \mathbf{v}_W = \mathbf{v}_M + {}_W\mathbf{v}_M$$

This is the standard result.

Rearrange to make $\mathbf{v}_W$ the subject, since this is the **same** in the two scenarios.

### Scenario 1

|  | Magnitude | Direction |
|---|---|---|
| $\mathbf{v}_W$ | $x$ | From N$\theta$°W |
| $\mathbf{v}_M$ | $u$ | due E |
| ${}_W\mathbf{v}_M$ | ? | From N10°W |

Both $x$ and $\theta$ will be the **same** in both scenarios and we are trying to find $\theta$.

Draw the vector triangle

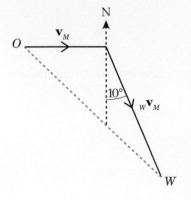

The resultant represents the velocity of the wind, in both magnitude and direction.

## Scenario 2

|  | Magnitude | Direction |
|---|---|---|
| $\mathbf{v}_W$ | $x$ | From N$\theta$°W |
| $\mathbf{v}_M$ | $u$ | due N |
| $_W\mathbf{v}_M$ | ? | From N30°W |

Draw the vector triangle

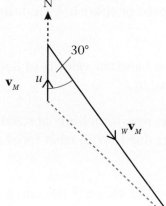

The resultant represents the velocity of the wind, in both magnitude and direction.

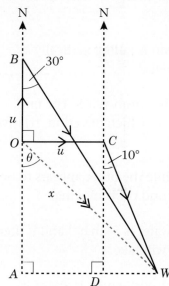

Since the resultant is the **same** for both triangles, we can put them together, as shown

$\overrightarrow{OW}$ represents the velocity of the wind.

The directions $AD$ and $AW$ are due E.

In $\triangle ABW$, $\tan 30° = \dfrac{x \sin \theta}{x \cos \theta + u}$

Although there are 3 unknowns here, we can solve these equations for $\theta$.

In $\triangle CDW$, $\tan 10° = \dfrac{x \sin \theta - u}{x \cos \theta}$

$x \cos \theta + u = x \sin \theta \cot 30°$

$x \sin \theta - u = x \cos \theta \tan 10°$

We first eliminate $u$ by adding these, then $x$ cancels.

$\cancel{x}(\cos \theta + \sin \theta) = \cancel{x}(\sin \theta \cot 30° + \cos \theta \tan 10°)$

$\sin \theta (1 - \cot 30°) = \cos \theta (\tan 10° - 1)$

$$\tan \theta = \dfrac{1 - \tan 10°}{\cot 30° - 1}$$

Solve for $\theta$.

$\theta = 48.4°$ (3 s.f.)

State the answer.

The wind blows from N48.4°W.

## Exercise 1A

**1** The velocity vectors of two particles $P$ and $Q$ are $\mathbf{v}_P$ and $\mathbf{v}_Q$ respectively. Find the velocity of $P$ relative to $Q$ and the relative speed of $Q$ to $P$ in each of the following cases:

**a** $\mathbf{v}_P = (5\mathbf{i} + 6\mathbf{j})\,\mathrm{m\,s^{-1}}$, $\qquad$ $\mathbf{v}_Q = (4\mathbf{i} - 3\mathbf{j})\,\mathrm{m\,s^{-1}}$

**b** $\mathbf{v}_P = 6\mathbf{j}\,\mathrm{m\,s^{-1}}$, $\qquad\qquad$ $\mathbf{v}_Q = (-2\mathbf{i} + \mathbf{j})\,\mathrm{m\,s^{-1}}$

**c** $\mathbf{v}_P = (5\mathbf{i} + 6\mathbf{j} - 3\mathbf{k})\,\mathrm{m\,s^{-1}}$, $\quad$ $\mathbf{v}_Q = (\mathbf{i} - 6\mathbf{i} + 2\mathbf{k})\,\mathrm{m\,s^{-1}}$.

**2** A man is driving due north at $40\,\mathrm{km\,h^{-1}}$ along a straight road when he notices that the wind appears to be coming from N60°W with a speed of $40\,\mathrm{km\,h^{-1}}$. Find the actual velocity of the wind.

**3** The velocity of $A$ relative to $B$ is $(2\mathbf{i} + 3\mathbf{j})\,\mathrm{m\,s^{-1}}$ and the velocity of $B$ relative to $C$ is $(-\mathbf{i} + 4\mathbf{j})\,\mathrm{m\,s^{-1}}$. Find the velocity of $A$ relative to $C$.

**4** A man who can row at $4\,\mathrm{km\,h^{-1}}$ in still water rows with his boat steering in the direction N20°E. There is a current of $2\,\mathrm{km\,h^{-1}}$ flowing due E. With what speed and in what direction does the boat actually move?

**5** A woman is walking along a road with a speed of $4\,\mathrm{km\,h^{-1}}$. The rain is falling vertically at $7\,\mathrm{km\,h^{-1}}$. At what angle to the vertical should she hold her umbrella?

**6** A bird can fly in still air at $100\,\mathrm{km\,h^{-1}}$. The wind blows at $90\,\mathrm{km\,h^{-1}}$ from W$\alpha$°S. The bird wishes to return to its nest which is due E of its present position. In which direction, relative to the air, should it fly?

**7** Two cars are moving at the same speed. The first is moving SE while the other appears to be approaching it from the east. Find the direction in which the second car is moving.

**8** A ship has to travel 20 km due E. If the speed of the ship in still water is $5\,\mathrm{km\,h^{-1}}$ and if there is a current of $3\,\mathrm{km\,h^{-1}}$ in the direction N30°E , find how long it will take.

**9** An aeroplane can fly at $600\,\mathrm{km\,h^{-1}}$ in still air. It has to fly to an airport which is SW of its current position. There is a wind of $90\,\mathrm{km\,h^{-1}}$ blowing from N20°W.

**a** What course should the aeroplane set?

**b** What is the ground speed of the aeroplane?

**10** A river flows at $2.5\,\mathrm{m\,s^{-1}}$. A fish swims from a point $P$ to a point $Q$, which is directly upstream from $P$, and then back to $P$ with speed $6.5\,\mathrm{m\,s^{-1}}$ relative to the water. A second fish, in the same time and with the same relative speed as the first fish, swims to the point $R$ on the bank directly opposite to $P$ and back to $P$. Find the ratio $PQ:PR$.

**11** A man is cruising in a boat which is capable of a speed of $10\,\mathrm{km\,h^{-1}}$ in still water. He is heading towards a marker buoy which is NE of his position and 6 km away. The current is running at a speed of $3\,\mathrm{km\,h^{-1}}$ due E.

**a** What course should he set?

**b** How long will it take to reach the buoy?

**12** A river flows at a speed $u$. A boat is rowed with speed $v$ relative to the river. The width of the river is $w$ and the boat is to reach the opposite bank at a distance $d$ downstream. Show that, if $\dfrac{uw}{\sqrt{w^2 + d^2}} < v < u$, there are two directions in which the boat may be steered.

**13** A car is moving due W and the wind appears, to the driver, to be coming from a direction N60°W. When he drives due E at the same speed the wind appears to be coming from a direction N30°E. If he now travels due S at the same speed, find the apparent direction of the wind.

**14** When a ship travels at $10\,\text{km h}^{-1}$ due N the wind appears to be coming from a direction N40°E. When the speed is increased to $25\,\text{km h}^{-1}$ the wind appears to be coming from a direction N25°E. Find the true speed and direction of the wind.

**15** A woman cycles due N at $40\,\text{km h}^{-1}$ and the wind seems to be blowing from the East. When she cycles due S at $50\,\text{km h}^{-1}$, the wind seems to be blowing from the South East. Find the true velocity of the wind.

**16** When a motorcyclist travels along a straight road at $48\,\text{km h}^{-1}$ due N, the wind seems to be blowing from a direction N40°E. When he returns along the same road at the same speed, the wind seems to be blowing from a direction S30°E. Find the true speed and direction of the wind.

## 1.4 You can solve problems which involve one body intercepting (or colliding with) another body.

For two moving people, $P$ and $Q$, the velocity of $P$ relative to $Q$ is how $Q$ would see $P$ moving. Thus, if the two people were on a collision course then $Q$ would see $P$ moving directly towards him at all times, until they collide. We can prove this as follows:

Suppose at $t = 0$, $P$ is at the point $P_0$ with position vector $\mathbf{r}_{P_0}$ and $Q$ is at the point $Q_0$ with position vector $\mathbf{r}_{Q_0}$.

Then at time t,

$$\mathbf{r}_P = \mathbf{r}_{P_0} + t\mathbf{v}_P \qquad \text{and}$$
$$\mathbf{r}_Q = \mathbf{r}_{Q_0} + t\mathbf{v}_Q.$$

■ If $P$ and $Q$ collide, there exists a $t$-value for which $\mathbf{r}_p = \mathbf{r}_Q$

   i.e. at this $t$ value, $(t > 0)$,

$\mathbf{r}_{P_0} + t\mathbf{v}_P = \mathbf{r}_{Q_0} + t\mathbf{v}_Q$

thus $\mathbf{r}_{Q_0} - \mathbf{r}_{P_0} = t(\mathbf{v}_P - \mathbf{v}_Q) = t{}_P\mathbf{v}_Q$

   i.e. $\dfrac{1}{t}\overrightarrow{P_0Q_0} = {}_P\mathbf{v}_Q \quad (\dfrac{1}{t} > 0)$

i.e. the velocity of $P$ relative to $Q$ is directed along the line **from** the **initial** position of $P$ **towards** the initial position of $Q$, since $\dfrac{1}{t} > 0$.

Note that $P$ and $Q$ are interchangeable in this argument.

**Fixing the target**

If $P$ is 'chasing' or 'attempting to intercept' $Q$, we can view the above in a slightly different way.

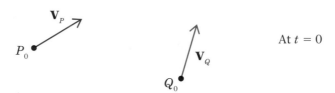

At $t = 0$

We 'fix the target' by applying the vector $(-\mathbf{v}_Q)$ to $Q$ but we must also apply this vector to $P$ as well. (This is equivalent to 'consider the motion relative to $Q$'.)

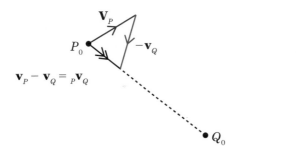

$\mathbf{v}_P - \mathbf{v}_Q = {}_P\mathbf{v}_Q$

$P$ is now moving with velocity
$\mathbf{v}_P - \mathbf{v}_Q = {}_P\mathbf{v}_Q$ and $Q$ is at rest.

$P$ will intercept $Q$ if ${}_P\mathbf{v}_Q$ is 'pointing towards $Q_0$', as before, i.e. ${}_P\mathbf{v}_Q$ is in the same direction as $\overrightarrow{P_0Q_0}$.

## Example 6

At 2 p.m. two ships $P$ and $Q$ are 10 km apart with $P$ due west of $Q$. The ship $P$ is travelling at $20\,\text{km h}^{-1}$ in a direction N60°E and ship $Q$ is travelling at $10\,\text{km h}^{-1}$ due N.

**a** Show that the two ships will collide.

**b** Find the time, to the nearest minute, when the collision occurs.

**a**

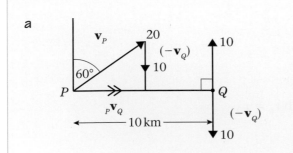

Draw a diagram showing the initial positions of $P$ and $Q$ and their velocities.

Fix $Q$, by applying a vector of magnitude 10 due S. This vector must then be applied to $P$.

Since $20 \cos 60° = 10$, ${}_P\mathbf{v}_Q$ is due E and a collision must take place.

Alternatively, we could fix $P$ i.e. consider the motion relative to $P$.

Since $Q$ is initially due E of $P$.

b  $t = \dfrac{PQ}{|_P\mathbf{v}_Q|} = \dfrac{10}{20 \sin 60°}$

$= \dfrac{1}{\sqrt{3}}$ h

$= 35$ minutes

Time of collision is 2.35 p.m.

time $= \dfrac{\text{initial separation}}{\text{relative speed}}$

State the answer.

## Example 7

At 7 p.m. two ships $S$ and $T$ are 20 km apart with $T$ on a bearing of 250° from $S$. Ship $S$ is moving at 4 km h$^{-1}$ on a bearing of 320°. The maximum speed of ship $T$ is 7 km h$^{-1}$.

**a** Find the course that $T$ should set in order to intercept $S$ as soon as possible.

**b** Find the time at which the interception occurs.

**c** Find the distance from the initial position of $S$ to the point of interception.

a

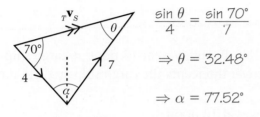

Draw a diagram showing the initial positions of $S$ and $T$ and the velocity of $S$. We then **fix** the target $S$.

Note that the vector diagram is separate from the displacement diagram, although we include both together.

The vector $\triangle$ is:

$\dfrac{\sin \theta}{4} = \dfrac{\sin 70°}{7}$

$\Rightarrow \theta = 32.48°$

$\Rightarrow \alpha = 77.52°$

Use the sine rule.

We need $\alpha$ to find the course that $T$ moves on.

∴ Course is 77.52° − 40°

$= 37.52°$

$= 038°$

Give answer as a 3-figure bearing.

**b** $|_T\mathbf{v}_S|^2 = 4^2 + 7^2 - 2 \times 4 \times 7 \cos \alpha$

$\qquad = 65 - 56 \cos \alpha$

$|_T\mathbf{v}_S| = 7.273 \text{ km h}^{-1}$

$\text{time} = \dfrac{20}{7.273} \text{ h}$

$\qquad = 2.75 \text{ h}$

i.e. time is 9.45 p.m.

**c** Distance travelled by S to the point of interception

$\qquad = 4 \times 2.75$

$\qquad = 11 \text{ km}$

> We need to find the relative speed of the ships.

> $\text{time} = \dfrac{\text{initial separation}}{\text{relative speed}}$

> Note we are using the **actual** speed of S.

## Exercise **1B**

**1** At 10.30 a.m. an aeroplane has position vector $(-100\mathbf{i} + 220\mathbf{j})$ km and is moving with constant velocity $(300\mathbf{i} + 400\mathbf{j})$ km h$^{-1}$. At 10.45 a.m. a cargo plane has position vector $(-60\mathbf{i} + 355\mathbf{j})$ km and is moving with constant velocity $(400\mathbf{i} + 300\mathbf{j})$ km h$^{-1}$.

   **a** Show that the planes will crash if they maintain these velocities.

   **b** Find the time at which the crash will occur.

   **c** Find the position vector of the point at which the crash takes place.

**2** Hiker $A$ is 3 km due W of hiker $B$. Hiker $A$ walks due N at 5 km h$^{-1}$. Hiker $B$ starts at the same time and walks at 7 km h$^{-1}$.

   **a** In what direction should $B$ walk in order to meet $A$?

   **b** How long will it take to do so?

**3** A batsman strikes a cricket ball at 15 m s$^{-1}$ on a bearing of 260°. A fielder is standing 45 m from the batsman on a bearing of 245°. He runs at 6 m s$^{-1}$ to intercept the ball.

   **a** Find the direction in which the fielder should run in order to intercept the ball as quickly as possible.

   **b** Find the time, to 1 decimal place, that it takes him to do so.

**4** A destroyer, moving at 48 km h$^{-1}$ in a direction N30°E, observes, at 12 noon, a cargo ship which is steaming due N at 20 km h$^{-1}$. The destroyer intercepts the cargo ship at 12.45 p.m. Find

   **a** the distance of the cargo ship from the destroyer at 12 noon,

   **b** the bearing of the cargo ship from the destroyer at 12 noon.

**5** A speedboat moving at 75 km h$^{-1}$ wishes to intercept a yacht which is moving at 20 km h$^{-1}$ in a direction 040°. Initially the speedboat is 10 km from the yacht on a bearing of 150°.

   **a** Find the course that the speedboat should set in order to intercept the yacht.

   **b** Find how long the journey will take.

**6** A lifeboat sets out from a harbour at 10.10 a.m. to go to the assistance of a dinghy which is, at that time, 5 km due S of the harbour and drifting at 8 km h$^{-1}$ due W. The lifeboat can travel at 20 km h$^{-1}$. Find the course that it should set in order to reach the yacht as quickly as possible and find the time when it arrives.

**7** A gunner in a bomber, which is flying N50°E at 320 m s$^{-1}$ wishes to fire at a fighter plane which is flying S60°W at 360 m s$^{-1}$. If the gun fires its shell at 1000 m s$^{-1}$, in what direction should the gun be aimed when the fighter is due E of the bomber?

## 1.5 You can solve problems which involve closest approach.

■ If a body, *A*, is 'chasing' another body, *B*, it may not be moving sufficiently fast to intercept the target but there will be a point at which *A* is closest to *B*, the **point of closest approach**.

### Example 8

At 3 p.m. the position vectors and velocity vectors of two ships, *P* and *Q*, are as follows:

$$\mathbf{r}_P = \begin{pmatrix} 2 \\ 1 \end{pmatrix} \text{km} \quad \mathbf{v}_P = \begin{pmatrix} 3 \\ 1 \end{pmatrix} \text{km h}^{-1} \quad \text{and} \quad \mathbf{r}_Q = \begin{pmatrix} -1 \\ -4 \end{pmatrix} \text{km} \quad \mathbf{v}_Q = \begin{pmatrix} 11 \\ 3 \end{pmatrix} \text{km h}^{-1}$$

**a** Assuming that these velocities are maintained, find the least distance between the ships in the subsequent motion.

**b** Find at what time this closest approach occurs.

**c** Find the position vectors of the two ships at this time.

**Method 1: Using differentiation**

**a** and **b**

At time *t*, $\mathbf{r}_P = \begin{pmatrix} 2 \\ 1 \end{pmatrix} + t\begin{pmatrix} 3 \\ 1 \end{pmatrix}$

We first find the position of each ship *t* hours after 3 p.m.

$$\mathbf{r}_Q = \begin{pmatrix} -1 \\ -4 \end{pmatrix} + t\begin{pmatrix} 11 \\ 3 \end{pmatrix}$$

$$\Rightarrow {}_P\mathbf{r}_Q = \mathbf{r}_P - \mathbf{r}_Q = \begin{pmatrix} 3 \\ 5 \end{pmatrix} + t\begin{pmatrix} -8 \\ -2 \end{pmatrix}$$

We then subtract to give the relative position vector of *P* from *Q* i.e. $\overrightarrow{QP}$ at time *t*.

$$= \begin{pmatrix} 3 - 8t \\ 5 - 2t \end{pmatrix}$$

So, $|{}_P\mathbf{r}_Q|^2 = (3 - 8t)^2 + (5 - 2t)^2 = X$ ①

Note that we do not need to find the square root of this expression.

then, $\frac{dX}{dt} = -16(3 - 8t) - 4(5 - 2t)$ ②

To minimise this expression, differentiate with respect to *t* and equate to zero.

For minimum X, $\frac{dX}{dt} = 0$

i.e. $-16(3 - 8t) - 4(5 - 2t) = 0$

$$12 - 32t + 5 - 2t = 0$$

$$17 = 34t$$

$$\tfrac{1}{2} = t$$

Solve for $t$.

$\therefore X_{MIN} = (3 - 4)^2 + (5 - 1)^2$

Substitute in $t = \tfrac{1}{2}$ in ①.

$\qquad = 1 + 16 = 17$

Minimum distance is $\sqrt{17}$ km at 3.30 p.m.

We must remember to find the square root here.

**c** $\quad \mathbf{r}_P = \begin{pmatrix} 2 \\ 1 \end{pmatrix} + \tfrac{1}{2}\begin{pmatrix} 3 \\ 1 \end{pmatrix}$

$\qquad = \begin{pmatrix} 3\tfrac{1}{2} \\ 1\tfrac{1}{2} \end{pmatrix}$ km

Substitute in $t = \tfrac{1}{2}$.

$\quad \mathbf{r}_Q = \begin{pmatrix} -1 \\ -4 \end{pmatrix} + \tfrac{1}{2}\begin{pmatrix} 11 \\ 3 \end{pmatrix}$

$\qquad = \begin{pmatrix} 4\tfrac{1}{2} \\ -2\tfrac{1}{2} \end{pmatrix}$ km

**Method 2: Using the scalar product and relative velocity**

Fix $Q$

Alternatively, fix $P$!

i.e. consider the motion relative to $Q$

Then $\ _P\mathbf{v}_Q = \mathbf{v}_P - \mathbf{v}_Q = \begin{pmatrix} 3 \\ 1 \end{pmatrix} - \begin{pmatrix} 11 \\ 3 \end{pmatrix}$

$\qquad\qquad = \begin{pmatrix} -8 \\ -2 \end{pmatrix}$

Alternatively see below.

As before, in Method 1, at time $t$ hours after 3 p.m.

$$_P\mathbf{r}_Q = \begin{pmatrix} 3 - 8t \\ 5 - 2t \end{pmatrix}$$

We could obtain $_P\mathbf{v}_Q$ by differentiating this vector with respect to time.

The value of $t$ at which closest approach occurs is when

$\qquad _P\mathbf{r}_Q$ and $_P\mathbf{v}_Q$ are **perpendicular**.

i.e. $\qquad\qquad _P\mathbf{r}_Q \cdot {_P\mathbf{v}_Q} = 0$

$$\begin{pmatrix} 3 - 8t \\ 5 - 2t \end{pmatrix} \cdot \begin{pmatrix} -8 \\ -2 \end{pmatrix} = 0$$

$-8(3 - 8t) - 2(5 - 2t) = 0$

$\qquad\qquad$ leading to $t = \tfrac{1}{2}$ as in Method 1.

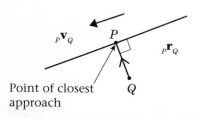

Point of closest approach

This equation is the same as ② in Method 1.

**Method 3: Using relative velocity and a diagram**

As before $_P\mathbf{v}_Q = \begin{pmatrix} -8 \\ -2 \end{pmatrix}$

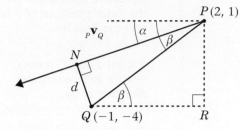

We are considering motion relative to Q.

We now need to draw **a reasonably accurate diagram** showing the initial positions of P and Q and the direction of the relative velocity $_P\mathbf{v}_Q$.

N is the point of closest approach

d is the **least separation.**

In $\triangle PQN$, $d = PQ \sin(\beta - \alpha)$

We now use trigonometry and coordinate geometry to find d.

$$PQ = \sqrt{(2 - -1)^2 + (1 - -4)^2}$$

$$= \sqrt{34} \text{ km}$$

$$\tan \beta = \frac{PR}{QR} = \frac{5}{3} \ (\beta = 59.04°)$$

$$\tan \alpha = \frac{2}{8} = \frac{1}{4} \ (\alpha = 14.04°)$$

We could find $\sin(\beta - \alpha)$ **exactly**
$= \sin \beta \cos \alpha - \cos \beta \sin \alpha$
$= \dfrac{5}{\sqrt{34}} \dfrac{4}{\sqrt{17}} - \dfrac{3}{\sqrt{34}} \cdot \dfrac{1}{\sqrt{17}} = \dfrac{17}{\sqrt{34}\sqrt{17}} = \dfrac{1}{\sqrt{2}}$

$$\therefore \quad d = \sqrt{34} \sin 45°$$

$$= \sqrt{17} \text{ km as before}$$

To find $t$:

$$t = \frac{PN}{|_P\mathbf{v}_\alpha|} = \frac{\sqrt{34} \cos(\beta - \alpha)}{\sqrt{8^2 + 2^2}}$$

This method is less suitable when the information about the positions and velocities of the bodies is given in **component form**.

$$= \frac{\sqrt{34} \times \frac{1}{\sqrt{2}}}{\sqrt{68}} = \sqrt{\frac{17}{68}}$$

$$= \tfrac{1}{2}$$

as before

## Example 9

Two aircraft P and Q are flying, at the same altitude, with velocities $200 \text{ m s}^{-1}$ on a bearing of 030° and $300 \text{ m s}^{-1}$ on a bearing of 310° respectively. At 12 noon the aircraft are 2 km apart with P on a bearing of 290° from Q. Given that P and Q maintain their velocities.

**a** find the shortest distance between them in the subsequent motion,

**b** find the time when they are closest.

Fix P

Consider motion relative to P.

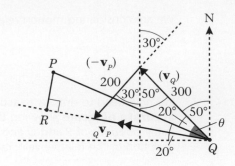

Draw a **reasonably accurate** diagram showing the initial positions of P and Q and the velocities.

P is fixed and Q moves, relative to it, along the direction of $_Q\mathbf{v}_P$.

R is the point of closest approach.

The closest approach is PR.

In the velocity △,

$$|_Q\mathbf{v}_P|^2 = 200^2 + 300^2 - 2 \times 200 \times 200 \times \cos 80°$$

Use cosine rule.

$$= 100^2(2^2 + 3^2 - 2 \times 2 \times 2 \times \cos 80°)$$

$$|_Q\mathbf{v}_P| = 100\sqrt{13 - 12\cos 80°}$$

Square root both sides.

$$\frac{\sin \theta}{200} = \frac{\sin 80°}{100\sqrt{13 - 12\cos 80°}}$$

Use sine rule.

$$\sin \theta = \frac{2\sin 80°}{\sqrt{13 - 12\cos 80°}}$$

$\theta$ must be acute since one of the other angles is 80°.

$$\Rightarrow \theta = 36.6°$$

$$PR = PQ \sin(\theta - 20°)$$

In △PQR, PQ = 2 km = 2000 m

$$= 2000 \sin 16.6°$$

$$\simeq 571\,\text{m (3 s.f.)}$$

$$t = \frac{QR}{|_Q\mathbf{v}_P|}$$

As P is 'fixed' and Q moves along QR with the relative speed.

$$= \frac{2000 \cos 16.6°}{100\sqrt{13 - 12\cos 80°}}$$

$$= 5.8\,\text{s (1 d.p.)}$$

They are closest 5.8 s after noon.

## Exercise 1C

**1** The position vectors and velocity vectors of two ships P and Q at 9 a.m. are as follows

$\mathbf{r}_P = (2\mathbf{i} + \mathbf{j})\,\text{km}$          $\mathbf{v}_P = (3\mathbf{i} + \mathbf{j})\,\text{km h}^{-1}$

$\mathbf{r}_Q = (-\mathbf{i} - 4\mathbf{j})\,\text{km}$          $\mathbf{v}_Q = (11\mathbf{i} + 3\mathbf{j})\,\text{km h}^{-1}$

Assuming that these velocities remain constant, find

  **a** the least distance between P and Q in the subsequent motion,

  **b** the time at which this least separation occurs.

**2** The position vectors and velocity vectors of two ships $P$ and $Q$ at certain times are as follows

$\mathbf{r}_P = (\mathbf{i} + 4\mathbf{j})\,\text{km}$          $\mathbf{v}_P = (4\mathbf{i} + 8\mathbf{j})\,\text{km h}^{-1}$      at 9 a.m.

$\mathbf{r}_Q = (20\mathbf{j})\,\text{km}$          $\mathbf{v}_Q = (9\mathbf{i} - 2\mathbf{j})\,\text{km h}^{-1}$      at 8 a.m.

Assuming that these velocities remain constant, find

**a** the least distance between $P$ and $Q$ in the subsequent motion,

**b** the time at which this least separation occurs.

**3** The position vectors and velocity vectors of two ships $P$ and $Q$ at certain times are as follows

$\mathbf{r}_P = (8\mathbf{i} - \mathbf{j})\,\text{km}$          $\mathbf{v}_P = (3\mathbf{i} + 7\mathbf{j})\,\text{km h}^{-1}$      at 3 p.m.

$\mathbf{r}_Q = (3\mathbf{i} + \mathbf{j})\,\text{km}$          $\mathbf{v}_Q = (2\mathbf{i} + 3\mathbf{j})\,\text{km h}^{-1}$      at 2 p.m.

Assuming that these velocities remain constant, find

**a** the least distance between $P$ and $Q$ in the subsequent motion,

**b** the time at which this least separation occurs.

**4** The position vectors and velocity vectors of two ships $P$ and $Q$ at 3 p.m. are as follows

$\mathbf{r}_P = (3\mathbf{i} - 5\mathbf{j})\,\text{km}$          $\mathbf{v}_P = (15\mathbf{i} + 14\mathbf{j})\,\text{km h}^{-1}$

$\mathbf{r}_Q = (13\mathbf{i} + 5\mathbf{j})\,\text{km}$          $\mathbf{v}_Q = (3\mathbf{i} - 10\mathbf{j})\,\text{km h}^{-1}$

Assuming that these velocities remain constant,

**a** find the least distance between $P$ and $Q$ in the subsequent motion.

Ship $Q$ has guns with a range of up to 5 km.

**b** Find the length of time for which ship $P$ is within the range of ship $Q$'s guns.

**5** The position vectors and velocity vectors of two ships $P$ and $Q$ at certain times are as follows

$\mathbf{r}_P = (-2\mathbf{i} + 3\mathbf{j})\,\text{km}$          $\mathbf{v}_P = (12\mathbf{i} - 4\mathbf{j})\,\text{km h}^{-1}$      at 2.45 p.m.

$\mathbf{r}_Q = (8\mathbf{i} + 7\mathbf{j})\,\text{km}$          $\mathbf{v}_Q = (2\mathbf{i} - 14\mathbf{j})\,\text{km h}^{-1}$      at 3 p.m.

Assuming that these velocities remain constant,

**a** find the least distance between $P$ and $Q$ in the subsequent motion.

Ship $Q$ has guns with a range of up to 2 km.

**b** Find the length of time for which ship $P$ is within the range of ship $Q$'s guns.

## Exercise 1D

**1** Two straight roads cross at right angles. A woman leaves the cross-roads and walks due E at 4.5 km h$^{-1}$. At the same time another woman leaves a point 10 km due S of the cross-roads and walks due N at 6 km h$^{-1}$.

**a** After how long will they be closest together?

**b** How far apart will they then be?

**2** Two trains are travelling on railway lines which cross at right angles. The first train is travelling at $45\,\text{km}\,\text{h}^{-1}$ and the second is travelling at $108\,\text{km}\,\text{h}^{-1}$.

**a** Find their relative speed.

The slower train passes the point where the lines cross one minute before the faster train.

**b** Find the shortest distance between the trains.

**3** At 10 a.m. an aircraft $A$ is $300\,\text{km}$ N50°E of another aircraft $B$. Aircraft $A$ is flying at $800\,\text{km}\,\text{h}^{-1}$ in the direction N70°W and aircraft $B$ is flying at $600\,\text{km}\,\text{h}^{-1}$ in the direction N10°W.

**a** Find the least distance between the aircraft in the subsequent motion.

**b** Find the time when they are closest to each other.

**4** A ship $P$ steams at $20\,\text{km}\,\text{h}^{-1}$ on a bearing of 015°. Another ship $Q$ steams at $12\,\text{km}\,\text{h}^{-1}$ on a bearing of 330°.

**a** Find the velocity of $Q$ relative to $P$.

At 12 noon $Q$ is $5\,\text{km}$ due E of $P$. If they maintain their velocities,

**b** find the shortest distance between the ships.

**5** At a particular instant a liner is $1\,\text{km}$ NW of a tanker. The liner is moving at $18\,\text{km}\,\text{h}^{-1}$ due E and the tanker is moving at $15\,\text{km}\,\text{h}^{-1}$ NE.

**a** Find the shortest distance between the ships.

**b** Find the interval of time that passes until they are at the point of closest approach.

---

**1.6** **You can solve problems which involve setting a course to achieve closest approach.**

### Example 10

A ship $P$ is moving on a bearing of 055° with speed $18\,\text{km}\,\text{h}^{-1}$. A second ship $Q$ is initially $6\,\text{km}$ from $P$ on a bearing of 155°, and moves with speed $15\,\text{km}\,\text{h}^{-1}$.

**a** Find the course that $Q$ must set in order to pass as close to $P$ as possible.

**b** Find the distance between the ships when they are closest.

**c** Find the time taken for $Q$ to reach this position.

**d** Find the distance travelled by $P$ to reach the position of closest approach.

Fix the target i.e. $P$

i.e. apply $-\mathbf{v}_P$ to both $P$ and $Q$.

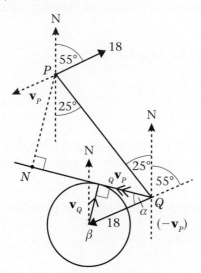

Consider the motion relative to $P$.

An accurate diagram showing the initial positions of the ships, together with their velocities, is essential.

The circle, of radius 15, shows all possible positions of $\mathbf{v}_Q$. The one which takes $Q$ **closest to $P$** (which is fixed) is the one which makes $_Q\mathbf{v}_P$ the tangent from $Q$ to this circle.

$$\sin \alpha = \frac{15}{18} \Rightarrow \alpha = 56.44°$$
$$\Rightarrow 90° - \alpha = 33.56°$$

$N$ is the point of closest approach.

**a**   Direction of $\mathbf{v}_Q$ is $(55° - 33.56°)$
    i.e. $021.4°$

See diagram.

**b**   $P\hat{Q}N = 180° - 55° - 25° - \alpha$
$$= 43.56°$$
so, $PN = PQ \sin 43.56°$
$$= 6 \sin 45.56°$$
$$= 4.135 \text{ km}$$

We now use $\triangle PQN$ to find $PN$.

**c**   time $= \dfrac{QN}{|_Q\mathbf{v}_P|}$

$$= \frac{PQ \cos 43.56°}{\sqrt{18^2 - 15^2}}$$

$$= \frac{6 \cos 43.56°}{\sqrt{99}}$$

$$= 0.437 \text{ hours}$$

$$\simeq 26 \text{ minutes}$$

Time to position of closest approach ($N$) is
$$\frac{QN}{\text{relative speed}}$$
Use Pythagoras to find $|_Q\mathbf{v}_P|$, the relative speed.

**d**   Distance travelled by $P = 18 \times 0.437$
$$= 7.866 \text{ km}$$
$$= 7.87 \text{ km} \quad (3 \text{ s.f.})$$

## Exercise 1E

1 $X$ and $Y$ are two yachts and $X$ is sailing at a constant speed of $15\,\text{km}\,\text{h}^{-1}$ in a direction N30°E. At 2 p.m. $Y$ is 4 km due E of $X$. Given that $Y$ travels at a constant speed of $12\,\text{km}\,\text{h}^{-1}$,

   **a** show that it is not possible for $Y$ to intercept $X$,

   **b** find the course that $Y$ should set in order to get as close as possible to $X$,

   **c** find the shortest distance between the yachts,

   **d** find the time when they are closest.

2 Two aircraft $P$ and $Q$ are flying at the same altitude. At 12 noon aircraft $Q$ is 5 km due S of aircraft $P$, and is flying at a constant $300\,\text{m}\,\text{s}^{-1}$ in the direction N60°E. If aircraft $P$ flies at a constant speed of $200\,\text{m}\,\text{s}^{-1}$, find

   **a** the direction in which it must fly in order to pass as close to aircraft $Q$ as possible,

   **b** the distance between the planes when they are closest,

   **c** the time when they are closest.

3 At 3 p.m. boat $C$ is due E of boat $B$ and $BC = 5.2\,\text{km}$. Boat $C$ is travelling due N at a constant speed of $13\,\text{km}\,\text{h}^{-1}$. Given that boat $B$ travels at $12\,\text{km}\,\text{h}^{-1}$, find

   **a** the course that $B$ should set in order to get as close as possible to $C$,

   **b** the shortest distance between the boats,

   **c** the time when this occurs,

   **d** the distance from the closest position of the boats to the initial position of $B$.

4 A fielder is placed at a distance of 30 m from a batsman and on a bearing of 250°. The batsman hits the ball at $17\,\text{m}\,\text{s}^{-1}$ in the direction N70°W. Given that the fielder runs at $8\,\text{m}\,\text{s}^{-1}$ from the moment the ball is struck, and ignoring any change in the speed of the ball, find

   **a** how close the fielder gets to the ball,

   **b** the time, from the instant when the ball was struck, that it takes the fielder to get to the closest position.

5 At 10 a.m. a frigate $F$ is 16 km due E of a cruiser $C$. The cruiser is moving at a constant speed of $40\,\text{km}\,\text{h}^{-1}$ on a bearing of 030° and the frigate is moving at a constant speed of $20\,\text{km}\,\text{h}^{-1}$. Find

   **a** the course that $F$ should set in order to get as close as possible to $C$,

   **b** the closest distance between them,

   **c** the time when this occurs.

   The guns on the frigate have a range of up to 10 km.

   **d** Find the length of time for which $C$ is within the range of ship $F$'s guns.

   The guns on the cruiser have a range of up to 9 km.

   **e** Find the length of time for which $F$ is within the range of ship $C$'s guns.

## Mixed exercise 1F

**1** Particles $P$, $Q$ and $R$ move in a plane with constant velocities. At time $t = 0$ the position vectors of $P$, $Q$ and $R$, relative to a fixed origin $O$, are $(\mathbf{i} + 3\mathbf{j})$ km, $(9\mathbf{i} + 9\mathbf{j})$ km and $(6\mathbf{i} + 13\mathbf{j})$ km respectively. The velocity of $R$ relative to $P$ is $(7\mathbf{i} - 10\mathbf{j})$ km h$^{-1}$ and the velocity of $R$ relative to $Q$ is $(9\mathbf{i} - 12\mathbf{j})$ km h$^{-1}$.

  **a** Find the velocity of $Q$ relative to $P$.

  **b** Show that $P$ and $Q$ do not collide.

  **c** Find the shortest distance between $P$ and $Q$.

  **d** Find the time taken to reach the position of closest approach.

  **e** Show that $Q$ and $R$ do collide.

  **f** Find the distance between $P$ and $R$ when this collision occurs.

**2** A ship is steaming due E at 10 km h$^{-1}$. A destroyer is 5 km due S of the ship and wishes to intercept it. If the destroyer can travel at 25 km h$^{-1}$,

  **a** in which direction will it travel,

  **b** how long will it take?

**3** Two trains $S$ and $T$ are moving at constant speed, $S$ at 50 km h$^{-1}$ NW and $T$ at a speed $v$ km h$^{-1}$ due W. If the velocity of $S$ relative $T$ is NE in direction,

  **a** show that it is 50 km h$^{-1}$ in magnitude,

  **b** find the value of $v$.

  If the speeds of $S$ and $T$ are interchanged,

  **c** find the velocity of $S$ relative to $T$ in magnitude and direction.

**4** A ship is travelling due E at 15 km h$^{-1}$ and is 10 km NW of a submarine. The submarine submerges immediately and travels at 10 km h$^{-1}$ NE underwater.

  **a** Show that when it crosses the ship's track, it is nearly 1 km behind.

  **b** Find the nearest distance to which it has approached the ship.

**5** A ship $A$ is moving at 14 km h$^{-1}$ due E and a ship $B$ is moving at 8 km h$^{-1}$ on a bearing of 040°. At 2 p.m., $A$ is 5 km due S of $B$. If the limit of visibility is 12 km, for how long after 2 p.m. is $B$ visible to $A$?

**6** A ship $P$ is steaming on a bearing of 225° at a constant speed of 8 km h$^{-1}$. A second ship $Q$ is sighted, 3 km SE of $P$, steaming due W at a constant speed. After a certain time, $Q$ is sighted 1 km due S of $P$. Find

  **a** the time taken, from the instant when $Q$ is first sighted, to the instant when $Q$ is due W of $P$,

  **b** the distance the ships are then apart,

  **c** the velocity of $Q$ relative to $P$.

**7** A side road running NW joins a main road which runs due N. Two cars, *A* and *B*, each travelling at 20 km h⁻¹, are approaching the junction between the two roads. At a particular instant, *A* is on the side road at a distance of 3 km from the junction and *B* is on the main road at a distance of 2 km from the junction. Given that the speeds of the cars remain constant, find

**a** how close to one another they get,

**b** the distance of *A* from the junction when this occurs.

**8** A ship is moving due W at 40 km h⁻¹ and the wind appears to blow from 67.5° west of south. The ship then steams due S at the same speed and the wind then appears to blow from 22.5° east of south. Find

**a** the true speed of the wind,

**b** the true direction of the wind.

**9** An aeroplane, which can fly at 160 km h⁻¹ in still air, starts from the point *A* to fly to the point *B* which is 400 km NE of *A*. If there is a wind of 40 km h⁻¹ blowing from the north, find

**a** the direction in which the aeroplane must fly,

**b** the time taken to reach *B*.

**10** A man can swim at a speed *u* relative to the water in a river which is flowing with speed *v*. Assuming that $u > v$, prove that it will take him $\dfrac{u}{\sqrt{u^2 - v^2}}$ times as long to swim a certain distance *d* upstream and back as it will to swim the same distance *d* and back in a direction perpendicular to the current, assuming that *d* is less than the width of the river.

# Summary of key points

**1** The position vector of $P$ relative to $Q$, $_P\mathbf{r}_Q$ is given by

$$_P\mathbf{r}_Q = \mathbf{r}_P - \mathbf{r}_Q,$$

where $\mathbf{r}_P$ and $\mathbf{r}_Q$ are the position vectors of $P$ and $Q$ relative to a fixed origin $O$.

**2** Two moving particles $P$ and $Q$ will collide if, at some time,

$$\mathbf{r}_P = \mathbf{r}_Q$$

i.e. $\quad _P\mathbf{r}_Q = \mathbf{0}$

**3** The velocity of $P$ relative to $Q$ ,$_P\mathbf{v}_Q$ is given by

$$_P\mathbf{v}_Q = \mathbf{v}_P - \mathbf{v}_Q$$

where $\mathbf{v}_P$ and $\mathbf{v}_Q$ are the actual velocities of $P$ and $Q$.

**4** For two particles $P$ and $Q$ to collide, $_P\mathbf{v}_Q = k\overrightarrow{P_0Q_0}$ where $k > 0$, and $P_0$ and $Q_0$ are the initial positions of $P$ and $Q$.

Note that the time to the collision is $\dfrac{1}{k}$.

**5** Two particles $P$ and $Q$ are closest together when $|_P\mathbf{r}_Q|$ has a minimum value.

This minimum value can be found by finding an expression for $|_P\mathbf{r}_Q|^2$ at time $t$, then differentiating it with respect to time and equating to zero.

**6** Alternatively, $|_P\mathbf{r}_Q|$ has a minimum value when $_P\mathbf{r}_Q$ and $_P\mathbf{v}_Q$ are perpendicular, i.e. when

$$_P\mathbf{r}_Q \cdot {_P\mathbf{v}_Q} = 0$$

**7** For $Q$ to pass as close to $P$ as possible, $\mathbf{v}_Q$ should be perpendicular to the path of $Q$ relative to $P$.

After completing this chapter you should be able to:
- solve problems about the impact of a smooth sphere with a fixed surface
- solve problems about the impact of smooth elastic spheres.

In this chapter you will find out how to use momentum, impulse and Newton's law of restitution to study collisions involving objects free to move in two dimensions.

# Elastic collisions in two dimensions

The first thing that many people think of when considering the impact of smooth elastic spheres is a sport such as snooker.

How does a player know how to play a shot off a side cushion?

### 2.1 You can solve problems about the oblique impact of a smooth sphere with a fixed surface.

- When a smooth sphere collides with a smooth flat surface and bounces off it the velocity of the sphere changes, and therefore the momentum changes. The change in momentum is caused by the impact between the sphere and the surface.

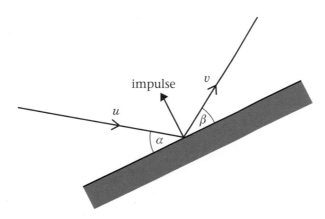

Because the sphere and the surface are smooth, we know that the reaction between the sphere and the surface acts along the common normal at the point of contact. This means that the impulse must act in the same direction. Consequently, in the impact between a smooth sphere and a fixed surface:

- **The impulse on the sphere acts perpendicular to the surface, through the centre of the sphere.**

- **Newton's law of restitution applies to the component of the velocity of the sphere perpendicular to the surface.**

- **The component of the velocity of the sphere parallel to the surface is unchanged.**

We can consider the motion parallel to the surface and the motion perpendicular to the surface separately. For the diagram above:

parallel to the plane there is no change: $v \cos \beta = u \cos \alpha$

perpendicular to the plane we can use Newton's law of restitution: $v \sin \beta = eu \sin \alpha$

Note that, by dividing,     $\tan \beta = e \tan \alpha$

and since $e \leqslant 1$     $\tan \beta < \tan \alpha$

so     $\beta < \alpha$

### Example 1

A smooth sphere $S$ is moving on a smooth horizontal plane with speed $u$ when it collides with a smooth fixed vertical wall. At the instant of collision the direction of motion of $S$ makes an angle of $60°$ with the wall. The coefficient of restitution between $S$ and the wall is $\frac{1}{4}$.

Find the speed of $S$ immediately after the collision.

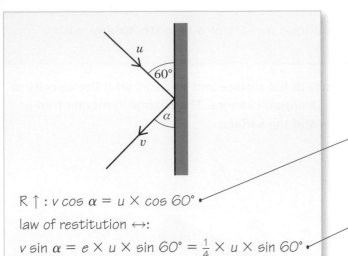

Start with a diagram to show what is happening.

The component of velocity parallel to the surface will be unchanged.

$R \uparrow : v \cos \alpha = u \times \cos 60°$

law of restitution $\leftrightarrow$:

$v \sin \alpha = e \times u \times \sin 60° = \frac{1}{4} \times u \times \sin 60°$

squaring and adding,

$v^2 = u^2(\frac{1}{4} + \frac{1}{16} \times \frac{3}{4}) = u^2 \times \frac{19}{64}, v = \frac{u\sqrt{19}}{8}$

Use Newton's law of restitution for motion perpendicular to the surface.

Substitute the values for $\cos 60°$ and $\sin 60°$ and use $\cos^2 \theta + \sin^2 \theta = 1$ to find $v$ in terms of $u$.

The same method can be used if the impact is with an inclined plane.

## Example 2

A small smooth ball is falling vertically. The ball strikes a smooth plane which is inclined at an angle $\alpha$ to the horizontal, where $\tan \alpha = \frac{1}{2}$. Immediately before striking the plane the ball has speed $5 \text{ m s}^{-1}$. The coefficient of restitution between the ball and the plane is $\frac{1}{2}$.

Find, to 3 significant figures, the speed of the ball immediately after the impact.

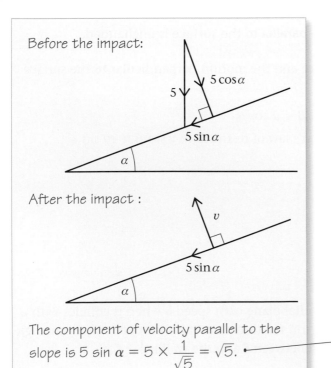

Start with a clear diagram showing the components of the velocity before and after the collision.

The component of velocity parallel to the slope is $5 \sin \alpha = 5 \times \frac{1}{\sqrt{5}} = \sqrt{5}$.

There is no change in the component parallel to the slope.

Perpendicular to the slope:

$$v = e \times 5 \cos \alpha = \frac{1}{2} \times 5 \times \frac{2}{\sqrt{5}} = \sqrt{5}$$

Therefore speed after impact

$$= \sqrt{\sqrt{5}^2 + \sqrt{5}^2} = \sqrt{10} \text{ m s}^{-1}.$$

> Newton's law of restitution applies perpendicular to the slope.

> Combine the two components using Pythagoras.

The problem could be described using vectors.

## Example 3

A small smooth ball of mass 2 kg is moving in the $xy$-plane and collides with a smooth fixed vertical wall which contains the $y$-axis. The velocity of the ball just before impact is $(-6\mathbf{i} - 4\mathbf{j}) \text{ m s}^{-1}$. The coefficient of restitution between the sphere and the wall is $\frac{1}{3}$. Find

**a** the speed of the ball immediately after the impact,

**b** the kinetic energy lost as a result of the impact.

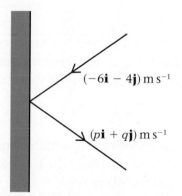

> Start with a diagram.

> Remember, $\mathbf{i}$ is a unit vector parallel to the $x$-axis, and $\mathbf{j}$ is a unit vector parallel to the $y$-axis.

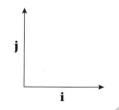

**a** Let the velocity of the ball after the impact be $(p\mathbf{i} + q\mathbf{j}) \text{ m s}^{-1}$.

Parallel to the wall : $q = -4$

Perpendicular to the wall:

$$p = \frac{1}{3} \times 6 = 2$$

The velocity of the ball after impact

$$= (2\mathbf{i} - 4\mathbf{j}) \text{ m s}^{-1}$$

∴ the speed of the ball is

$$\sqrt{2^2 + 4^2} = \sqrt{20} = 2\sqrt{5} \text{ m s}^{-1}.$$

**b** K.E. of the ball before impact

$$= \frac{1}{2} \times 2 \times (6^2 + 4^2) = 52$$

K.E. of the ball after impact

$$= \frac{1}{2} \times 2 \times 20 = 20$$

Loss of K.E. $= 52 - 20 = 32$ J

> The initial velocity is already in component form with one component parallel to the wall and the other perpendicular to it.

> Speed $= |\mathbf{v}|$, so use Pythagoras.

> Use K.E. $= \frac{1}{2}mv^2$ to find the kinetic energy just before and just after the impact.

> It is important to remember that in all of our examples, we are modelling the moving object as a particle, and the surfaces as smooth. Our model does not take any account of the possibility of any friction or that the object could be rolling or spinning. The answers we have found give an approximation to what actually happens.

## Exercise 2A

1. A smooth sphere $S$ is moving on a smooth horizontal plane with speed $u$ when it collides with a smooth fixed vertical wall. At the instant of collision the direction of motion of $S$ makes an angle of $\tan^{-1}\frac{3}{4}$ with the wall. The coefficient of restitution between $S$ and the wall is $\frac{1}{3}$.

   Find the speed of $S$ immediately after the collision.

2. A smooth sphere $S$ is moving on a smooth horizontal plane with speed $u$ when it collides with a smooth fixed vertical wall. At the instant of collision the direction of motion of $S$ makes an angle of $30°$ with the wall. Immediately after the collision the speed of $S$ is $\frac{7}{8}u$.

   Find the coefficient of restitution between $S$ and the wall.

3. A smooth sphere $S$ is moving on a smooth horizontal plane with speed $u$ when it collides with a smooth fixed vertical wall. At the instant of collision the direction of motion of $S$ makes an angle of $\tan^{-1}\frac{5}{12}$ with the wall. The coefficient of restitution between $S$ and the wall is $\frac{3}{5}$.

   Find the speed of $S$ immediately after the collision.

4. A smooth sphere $S$ is moving on a smooth horizontal plane with speed $u$ when it collides with a smooth fixed vertical wall. At the instant of collision the direction of motion of $S$ makes an angle of $\tan^{-1}2$ with the wall. Immediately after the collision the speed of $S$ is $\frac{3}{4}u$.

   Find the coefficient of restitution between $S$ and the wall.

5. A small smooth ball is falling vertically. The ball strikes a smooth plane which is inclined at an angle of $30°$ to the horizontal. Immediately before striking the plane the ball has speed $8\,\mathrm{m\,s^{-1}}$. The coefficient of restitution between the ball and the plane is $\frac{1}{4}$.

   Find the exact value of the speed of the ball immediately after the impact.

6. A small smooth ball is falling vertically. The ball strikes a smooth plane which is inclined at an angle of $20°$ to the horizontal. Immediately before striking the plane the ball has speed $10\,\mathrm{m\,s^{-1}}$. The coefficient of restitution between the ball and the plane is $\frac{2}{5}$. Find the speed, to 3 significant figures, of the ball immediately after the impact.

7. A small smooth ball of mass $750\,\mathrm{g}$ is falling vertically. The ball strikes a smooth plane which is inclined at an angle of $45°$ to the horizontal. Immediately before striking the plane the ball has speed $5\sqrt{2}\,\mathrm{m\,s^{-1}}$. The coefficient of restitution between the ball and the plane is $\frac{1}{2}$. Find

   a the speed, to 3 significant figures, of the ball immediately after the impact,

   b the magnitude of the impulse received by the ball as it strikes the plane.

8. A small smooth ball is falling vertically. The ball strikes a smooth plane which is inclined at an angle $\alpha$ to the horizontal, where $\tan \alpha = \frac{3}{4}$. Immediately before striking the plane the ball has speed $7.5\,\mathrm{m\,s^{-1}}$. Immediately after the impact the ball has speed $5\,\mathrm{m\,s^{-1}}$.

   Find the coefficient of restitution, to 2 significant figures, between the ball and the plane.

**9** A small smooth ball of mass 800 g is moving in the $xy$-plane and collides with a smooth fixed vertical wall which contains the $y$-axis. The velocity of the ball just before impact is $(5\mathbf{i} - 3\mathbf{j})\,\mathrm{m\,s^{-1}}$. The coefficient of restitution between the sphere and the wall is $\frac{1}{2}$. Find

  **a** the velocity of the ball immediately after the impact,

  **b** the kinetic energy lost as a result of the impact.

**10** A small smooth ball of mass 1 kg is moving in the $xy$-plane and collides with a smooth fixed vertical wall which contains the $x$-axis. The velocity of the ball just before impact is $(3\mathbf{i} + 6\mathbf{j})\,\mathrm{m\,s^{-1}}$. The coefficient of restitution between the sphere and the wall is $\frac{1}{3}$. Find

  **a** the speed of the ball immediately after the impact,

  **b** the kinetic energy lost as a result of the impact.

**11** A small smooth ball of mass 2 kg is moving in the $xy$-plane and collides with a smooth fixed vertical wall which contains the line $y = x$. The velocity of the ball just before impact is $(4\mathbf{i} + 2\mathbf{j})\,\mathrm{m\,s^{-1}}$. The coefficient of restitution between the sphere and the wall is $\frac{1}{3}$. Find

  **a** the velocity of the ball immediately after the impact,

  **b** the kinetic energy lost as a result of the impact.

**12** A smooth snooker ball strikes a smooth cushion with speed $8\,\mathrm{m\,s^{-1}}$ at an angle of 45° to the cushion. Given that the coefficient of restitution between the ball and the cushion is $\frac{2}{5}$, find the magnitude and direction of the velocity of the ball after the impact.

**13** A smooth snooker ball strikes a smooth cushion with speed $u\,\mathrm{m\,s^{-1}}$ at an angle of 50° to the cushion. The coefficient of restitution between the ball and the cushion is $e$.

  **a** Show that the angle between the cushion and the rebound direction is independent of $u$.

  **b** Find the value of $e$ given that the ball rebounds at right angles to its original direction.

**14** A smooth billiard ball strikes a smooth cushion at an angle of $\tan^{-1}\frac{3}{4}$ to the cushion. The ball rebounds at an angle of $\tan^{-1}\frac{5}{12}$ to the cushion. Find

  **a** the fraction of the kinetic energy of the ball lost in the collision,

  **b** the coefficient of restitution between the ball and the cushion.

**15** Two vertical walls meet at right angles at the corner of a room. A small smooth disc slides across the floor and bounces off each wall in turn. Just before the first impact the disc is moving with speed $u\,\mathrm{m\,s^{-1}}$ at an acute angle $\alpha$ to the wall. The coefficient of restitution between the disc and the wall is $e$. Find

  **a** the direction of the motion of the disc after the second collision,

  **b** the speed of the disc after the second collision.

**16** A small smooth sphere of mass $m$ is moving with velocity $(5\mathbf{i} - 2\mathbf{j})\,\mathrm{m\,s^{-1}}$ when it hits a smooth wall. It rebounds from the wall with velocity $(2\mathbf{i} + 2\mathbf{j})\,\mathrm{m\,s^{-1}}$. Find

  **a** the magnitude and direction of the impulse received by the sphere,

  **b** the coefficient of restitution between the sphere and the wall.

**17** A small smooth sphere of mass 2 kg is moving with velocity $(2\mathbf{i} + 3\mathbf{j})\,\mathrm{m\,s^{-1}}$ when it hits a smooth wall. It rebounds from the wall with velocity $(3\mathbf{i} - \mathbf{j})\,\mathrm{m\,s^{-1}}$. Find

   **a** the magnitude and direction of the impulse received by the sphere,

   **b** the coefficient of restitution between the sphere and the wall,

   **c** the kinetic energy lost by the sphere in the collision.

**18** Two smooth vertical walls stand on a smooth horizontal floor and intersect at an angle of $30°$. A particle is projected along the floor with speed $u\,\mathrm{m\,s^{-1}}$ at $45°$ to one of the walls and towards the intersection of the walls. The coefficient of restitution between the particle and each wall is $\dfrac{1}{\sqrt{3}}$. Find the speed of the particle after one impact with each wall.

---

**2.2** **You can solve problems about the oblique impact of two smooth spheres.**

The methods used to solve problems about the impact between a sphere and a fixed surface can be adapted to solve problems about the impact of two spheres.

At the moment of impact the two spheres have a common tangent, which is perpendicular to the line through the centres of the two spheres.

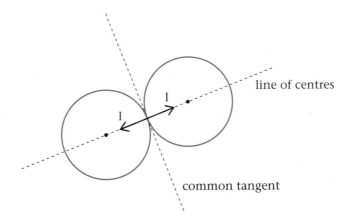

- **The reaction between the two spheres acts along the line of centres, so the impulse affecting each sphere also acts along the line of centres.**

- **The components of the velocities of the spheres perpendicular to the line of centres are unchanged in the impact.**

- **Newton's law of restitution applies to the components of the velocities of the spheres parallel to the line of centres.**

- **The law of conservation of momentum applies parallel to the line of centres.**

Of course the law of conservation of momentum also applies perpendicular to the line of centres, but trivially so since the components of velocity in this direction are unchanged.

## Example 4

A smooth sphere $A$, of mass 2 kg and moving with speed $6 \text{ m s}^{-1}$ collides obliquely with a smooth sphere $B$ of mass 4 kg. Just before the impact $B$ is stationary and the velocity of $A$ makes an angle of $60°$ with the lines of centres of the two spheres. The coefficient of restitution between the spheres is $\frac{1}{4}$. Find the magnitude and direction of the velocities of $A$ and $B$ immediately after the impact.

Before

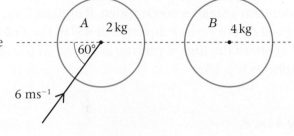

After

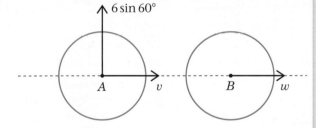

Start with a diagram and identify all the components that you need to find.

Since $B$ is stationary before the impact it will be moving along the line of centres after the impact.

No change perpendicular to the line of centres.

$A(\uparrow)$ Speed after collision $= 6 \sin 60° = 3\sqrt{3}$

Conservation of momentum.

Parallel to the line of centres:

$$2 \times 6 \cos 60° = 6 = 2v + 4w$$

Newton's law of restitution.

$$w - v = \frac{1}{4} \times 6 \cos 60° = \frac{6}{8} = \frac{3}{4}$$

$$\Rightarrow 2v + 4(v + \tfrac{3}{4}) = 6v + 3 = 6, \ v = \frac{1}{2}$$

Solve the simultaneous equations for $v$ and $w$.

$$w = \frac{3}{4} + \frac{1}{2} = \frac{5}{4}$$

Speed of $A$ is $\sqrt{(3\sqrt{3})^2 + (\tfrac{1}{2})^2} = \sqrt{27\tfrac{1}{4}} \text{ m s}^{-1}$ at

Combine the two components of the velocity of $A$.

$$\tan^{-1}\left(\frac{3\sqrt{3}}{\frac{1}{2}}\right) = 84.5° \text{ to the line of centres, and}$$

speed of $B$ is $1\tfrac{1}{4} \text{ m s}^{-1}$ along the line of centres.

Example 5

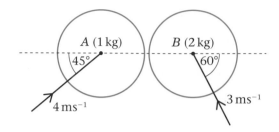

A small smooth sphere $A$ of mass 1 kg collides with a small smooth sphere $B$ of mass 2 kg. Just before the impact $A$ is moving with a speed of $4 \, \text{m s}^{-1}$ in a direction at 45° to the line of centres and $B$ is moving with speed $3 \, \text{m s}^{-1}$ at 60° to the line of centres, as shown in the diagram. The coefficient of restitution between the spheres is $\frac{3}{4}$. Find

**a** the kinetic energy lost in the impact,

**b** the magnitude of the impulse exerted by $A$ on $B$.

**a**

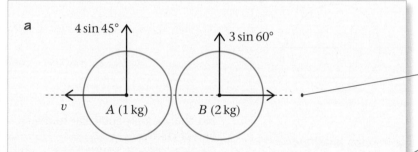

No change in the components perpendicular to the line of centres.

Conservation of momentum.

Newton's law of restitution.

Parallel to the line of centres:

$1 \times 4 \cos 45° - 2 \times 3 \cos 60° = 2w - v$

$2\sqrt{2} - 3 = 2w - v$

$v + w = \frac{3}{4}(4 \cos 45° + 3 \cos 60°)$

Solve the simultaneous equations to find $v$ and $w$.

$\frac{3\sqrt{2}}{2} + \frac{9}{8} = v + w, \quad 3w = \frac{7\sqrt{2}}{2} - \frac{15}{8}$

$w = \frac{7\sqrt{2}}{6} - \frac{5}{8} \approx 1.0249\ldots, \quad v = \frac{\sqrt{2}}{3} + \frac{7}{4} \approx 2.2214\ldots$

The loss in K.E. is due to the changes in the components of the velocities parallel to the line of centres.

K.E. lost by $A = \frac{1}{2} \times 1 \times \left((2\sqrt{2})^2 - 2.2214^2\right) = 1.53 \, \text{J}$

K.E. lost by $B = \frac{1}{2} \times 2 \times (1.5^2 - 1.0249^2) = 1.20 \, \text{J}$

Total K.E. lost = 2.73 J

Change in momentum along the line of centres – don't forget that the direction changed in the impact.

**b** Impulse on $B = 2(1.0249 + 1.5) = 5.05 \, \text{Ns}$

## Example 6

A smooth sphere $A$ of mass 5 kg is moving on a smooth horizontal surface with velocity $(2\mathbf{i} + 3\mathbf{j})\,\mathrm{m\,s^{-1}}$. Another smooth sphere $B$ of mass 3 kg and the same radius as $A$ is moving on the same surface with velocity $(4\mathbf{i} - 2\mathbf{j})\,\mathrm{m\,s^{-1}}$. The spheres collide when their line of centres is parallel to $\mathbf{j}$. The coefficient of restitution between the spheres is $\frac{3}{5}$. Find the velocities of both spheres after the impact.

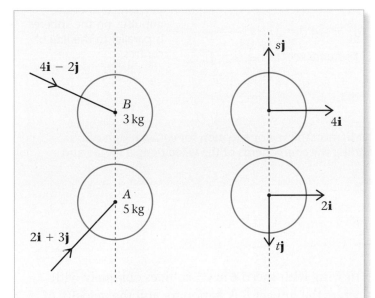

The diagram helps you to understand the relative positions of $A$ and $B$ if they are to collide – if they were the other way round they would be moving apart.

There is no change in the components of velocity perpendicular to the line of centres.

Parallel to the line of centres:

$3 \times -2 + 5 \times 3 = 3 \times s - 5 \times t, 9 = 3s - 5t$ — Conservation of momentum.

$s + t = \frac{3}{5}(2 + 3),\ s + t = 3,\ 3s + 3t = 9$ — Newton's law of restitution.

$\Rightarrow t = 0,\ s = 3$ — Solve for $s$ and $t$.

Velocity of $A$ is $2\mathbf{i}\,\mathrm{m\,s^{-1}}$, and

velocity of $B$ is $(4\mathbf{i} + 3\mathbf{j})\,\mathrm{m\,s^{-1}}$.

## Example 7

Two small smooth spheres $A$ and $B$ have equal radii. The mass of $A$ is $2m$ kg and the mass of $B$ is $3m$ kg. The spheres are moving on a smooth horizontal plane and they collide. Immediately before the collision the velocity of $A$ is $(5\mathbf{j})\,\mathrm{m\,s^{-1}}$ and the velocity of $B$ is $(3\mathbf{i} - \mathbf{j})\,\mathrm{m\,s^{-1}}$. Immediately after the collision the velocity of $A$ is $(3\mathbf{i} + 2\mathbf{j})\,\mathrm{m\,s^{-1}}$. Find

**a** the speed of $B$ immediately after the collision,

**b** a unit vector parallel to the line of centres of the spheres at the instant of the collision.

**a** If the velocity of B after the collision is **v** then

$$2m(5\mathbf{j}) + 3m(3\mathbf{i} - \mathbf{j}) = 2m(3\mathbf{i} + 2\mathbf{j}) + 3m\mathbf{v}$$

Forming a vector equation for conservation of momentum.

$$3\mathbf{v} = (9 - 6)\mathbf{i} + (10 - 3 - 4)\mathbf{j} = 3\mathbf{i} + 3\mathbf{j}, \mathbf{v} = (\mathbf{i} + \mathbf{j})\,\text{m s}^{-1}$$

Speed of $B = \sqrt{1^2 + 1^2} = \sqrt{2}\,\text{m s}^{-1}$

Speed $= |\mathbf{v}|$

**b** In the collision A receives an impulse of

$$2m(\,(3\mathbf{i} + 2\mathbf{j}) - (5\mathbf{j})) = 2m(3\mathbf{i} - 3\mathbf{j}) = 6m(\mathbf{i} - \mathbf{j})\,\text{Ns}$$

$\Rightarrow$ the line of centres is parallel to the unit vector $\dfrac{1}{\sqrt{2}}(\mathbf{i} - \mathbf{j})$

The direction of the impulses on the spheres is parallel to the line of centres.

Note that in this example it is much simpler to form the vector equation for conservation of momentum than the alternative route of finding the components of the velocities parallel to and perpendicular to the line of centres.

### Exercise 2B

**1** A smooth sphere $A$, of mass $2\,\text{kg}$ and moving with speed $6\,\text{m s}^{-1}$ collides obliquely with a smooth sphere $B$ of mass $4\,\text{kg}$. Just before the impact $B$ is stationary and the velocity of $A$ makes an angle of $10°$ with the lines of centres of the two spheres. The coefficient of restitution between the spheres is $\frac{1}{2}$. Find the magnitudes and directions of the velocities of $A$ and $B$ immediately after the impact.

**2** A smooth sphere $A$, of mass $4\,\text{kg}$ and moving with speed $4\,\text{m s}^{-1}$ collides obliquely with a smooth sphere $B$ of mass $2\,\text{kg}$. Just before the impact $B$ is stationary and the velocity of $A$ makes an angle of $30°$ with the lines of centres of the two spheres. The coefficient of restitution between the spheres is $\frac{1}{3}$. Find the magnitudes and directions of the velocities of $A$ and $B$ immediately after the impact.

**3** A smooth sphere $A$, of mass $3\,\text{kg}$ and moving with speed $5\,\text{m s}^{-1}$ collides obliquely with a smooth sphere $B$ of mass $4\,\text{kg}$. Just before the impact $B$ is stationary and the velocity of $A$ makes an angle of $45°$ with the lines of centres of the two spheres. The coefficient of restitution between the spheres is $\frac{1}{2}$. Find the magnitudes and directions of the velocities of $A$ and $B$ immediately after the impact.

**4** A small smooth sphere $A$ of mass $m$ and a small smooth sphere $B$ of the same radius but mass $2m$ collide. At the instant of impact, $B$ is stationary and the velocity of $A$ makes an angle $\theta$ with the line of centres. The direction of motion of $A$ is turned through $90°$ by the impact. The coefficient of restitution between the spheres is $e$. Show that

$$\tan^2 \theta = \frac{2e - 1}{3}.$$

**5** Two smooth spheres $A$ and $B$ are identical and are moving with equal speeds on a smooth horizontal surface. In the instant before impact, $A$ is moving in a direction perpendicular to the line of centres of the spheres, and $B$ is moving along the line of centres, as shown in the diagram. The coefficient of restitution between the spheres is $\frac{2}{3}$. Find the speeds and directions of motion of the spheres after the collision.

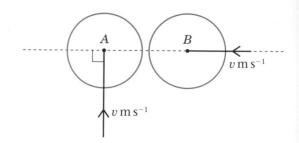

**6** A smooth sphere $A$ collides obliquely with an identical smooth sphere $B$. Just before the impact $B$ is stationary and the velocity of $A$ makes an angle of $\alpha$ with the line of centres of the two spheres. The coefficient of restitution between the spheres is $e\,(e \neq 1)$. Immediately after the collision the velocity of $A$ makes an angle of $\beta$ with the line of centres.

**a** Show that $\tan \beta = \dfrac{2 \tan \alpha}{1 - e}$.

**b** Hence show that in the collision the direction of motion of $A$ turns through an angle equal to $\tan^{-1}\!\left( \dfrac{(1 + e) \tan \alpha}{2 \tan^2 \alpha + 1 - e} \right)$.

**7**

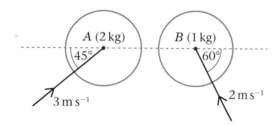

A small smooth sphere $A$ of mass 2 kg collides with a small smooth sphere $B$ of mass 1 kg. Just before the impact $A$ is moving with a speed of $3 \, \text{m s}^{-1}$ in a direction at $45°$ to the line of centres and $B$ is moving with speed $2 \, \text{m s}^{-1}$ at $60°$ to the line of centres, as shown in the diagram. The coefficient of restitution between the spheres is $\dfrac{\sqrt{2}}{3}$. Find

**a** the kinetic energy lost in the impact,

**b** the magnitude of the impulse exerted by $A$ on $B$.

**8** A small smooth sphere $A$ collides with an identical small smooth sphere $B$. Just before the impact $A$ is moving with a speed of $3\sqrt{2} \, \text{m s}^{-1}$ in a direction at $45°$ to the line of centres and $B$ is moving with speed $5 \, \text{m s}^{-1}$ at $\tan^{-1} \frac{4}{3}$ to the line of centres, as shown in the diagram. The coefficient of restitution between the spheres is $\frac{2}{3}$. Find

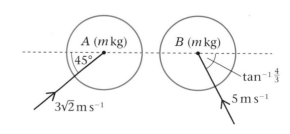

**a** the speeds of both spheres immediately after the impact,

**b** the fraction of the kinetic energy lost in the impact.

9   A smooth sphere $A$ of mass 2 kg is moving on a smooth horizontal surface with velocity $(4\mathbf{i} - \mathbf{j})\,\mathrm{m\,s^{-1}}$. Another smooth sphere $B$ of mass 4 kg and the same radius as $A$ is moving on the same surface with velocity $(2\mathbf{i} + \mathbf{j})\,\mathrm{m\,s^{-1}}$. The spheres collide when their line of centres is parallel to $\mathbf{j}$. The coefficient of restitution between the spheres is $\frac{1}{2}$. Find the velocities of both spheres after the impact.

10   A smooth sphere $A$ of mass 3 kg is moving on a smooth horizontal surface with velocity $(\mathbf{i} + 3\mathbf{j})\,\mathrm{m\,s^{-1}}$. Another smooth sphere $B$ of mass 1 kg and the same radius as $A$ is moving on the same surface with velocity $(-4\mathbf{i} + 2\mathbf{j})\,\mathrm{m\,s^{-1}}$. The spheres collide when their line of centres is parallel to $\mathbf{i}$. The coefficient of restitution between the spheres is $\frac{3}{4}$. Find the speeds of both spheres after the impact.

11   A smooth sphere $A$ of mass 1 kg is moving on a smooth horizontal surface with velocity $(2\mathbf{i} + 3\mathbf{j})\,\mathrm{m\,s^{-1}}$. Another smooth sphere $B$ of mass 2 kg and the same radius as $A$ is moving on the same surface with velocity $(-\mathbf{i} + \mathbf{j})\,\mathrm{m\,s^{-1}}$. The spheres collide when their line of centres is parallel to $\mathbf{i}$. The coefficient of restitution between the spheres is $\frac{3}{5}$. Find the kinetic energy lost in the impact.

12   Two small smooth spheres $A$ and $B$ have equal radii. The mass of $A$ is $m$ kg and the mass of $B$ is $2m$ kg. The spheres are moving on a smooth horizontal plane and they collide. Immediately before the collision the velocity of $A$ is $(2\mathbf{i} + 5\mathbf{j})\,\mathrm{m\,s^{-1}}$ and the velocity of $B$ is $(3\mathbf{i} - \mathbf{j})\,\mathrm{m\,s^{-1}}$. Immediately after the collision the velocity of $A$ is $(3\mathbf{i} + 2\mathbf{j})\,\mathrm{m\,s^{-1}}$. Find

  **a**  the velocity of $B$ immediately after the collision,

  **b**  a unit vector parallel to the line of centres of the spheres at the instant of the collision.

13   Two small smooth spheres $A$ and $B$ have equal radii. The mass of $A$ is $3m$ kg and the mass of $B$ is $m$ kg. The spheres are moving on a smooth horizontal plane and they collide. Immediately before the collision the velocity of $A$ is $(3\mathbf{i} - 5\mathbf{j})\,\mathrm{m\,s^{-1}}$ and the velocity of $B$ is $(4\mathbf{i} + \mathbf{j})\,\mathrm{m\,s^{-1}}$. Immediately after the collision the velocity of $A$ is $(4\mathbf{i} - 4\mathbf{j})\,\mathrm{m\,s^{-1}}$. Find

  **a**  the speed of $B$ immediately after the collision,

  **b**  the kinetic energy lost in the collision.

14   Two small smooth spheres $A$ and $B$ have equal radii. The mass of $A$ is $2m$ kg and the mass of $B$ is $m$ kg. The spheres are moving on a smooth horizontal plane and they collide. Immediately before the collision the velocity of $A$ is $(2\mathbf{i} + 5\mathbf{j})\,\mathrm{m\,s^{-1}}$ and the velocity of $B$ is $(2\mathbf{i} - 2\mathbf{j})\,\mathrm{m\,s^{-1}}$. Immediately after the collision the velocity of $A$ is $(3\mathbf{i} + 4\mathbf{j})\,\mathrm{m\,s^{-1}}$. Find

  **a**  the velocity of $B$ immediately after the collision,

  **b**  the coefficient of restitution between the two spheres.

15   A smooth uniform sphere $A$, moving on a smooth horizontal table, collides with an identical sphere $B$ which is at rest on the table. When the spheres collide the line joining their centres makes an angle of $45°$ with the direction of motion of $A$, as shown in the diagram. The coefficient of restitution between the spheres is $e$. The direction of motion of $A$ is deflected through an angle $\theta$ by the collision.

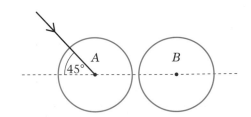

  Show that $\tan\theta = \dfrac{1 + e}{3 - e}$.

**16**

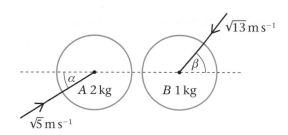

Two smooth uniform spheres $A$ and $B$ of equal radius have masses 2 kg and 1 kg respectively. They are moving on a smooth horizontal plane when they collide. Immediately before the collision the speed of $A$ is $\sqrt{5}$ m s$^{-1}$ and the speed of $B$ is $\sqrt{13}$ m s$^{-1}$. When they collide the line joining their centres makes an angle $\alpha$ with the direction of motion of $A$ and an angle $\beta$ with the direction of motion of $B$, where $\tan \alpha = \frac{1}{2}$ and $\tan \beta = \frac{3}{2}$, as shown in the diagram above. The coefficient of restitution between $A$ and $B$ is $\frac{1}{2}$.

Find the speed of each sphere after the collision.

**17**

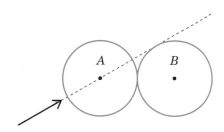

A smooth uniform sphere $B$ is at rest on a smooth horizontal plane, when it is struck by an identical sphere $A$ moving on the plane. Immediately before the impact, the line of motion of the centre of $A$ is tangential to the sphere $B$, as shown in the diagram above. The coefficient of restitution between the spheres is $\frac{1}{2}$. The direction of motion of $A$ is turned through an angle $\theta$ by the impact.

Show that $\tan \theta = \dfrac{3\sqrt{3}}{7}$.

## Mixed exercise 2C

**1** A smooth sphere $S$ is moving on a smooth horizontal plane with speed $u$ when it collides with a smooth fixed vertical wall. At the instant of collision the direction of motion of $S$ makes an angle of 45° with the wall. Immediately after the collision the speed of $S$ is $\frac{4}{5}u$.

Find the coefficient of restitution between $S$ and the wall.

**2** A small smooth ball of mass $\frac{1}{2}$ kg is falling vertically. The ball strikes a smooth plane which is inclined at an angle $\alpha$ to the horizontal, where $\tan \alpha = \frac{5}{12}$. Immediately before striking the plane the ball has speed 5.2 m s$^{-1}$. The coefficient of restitution between the ball and the plane is $\frac{1}{4}$. Find

**a** the speed, to 3 significant figures, of the ball immediately after the impact,

**b** the magnitude of the impulse received by the ball as it strikes the plane.

**3** A small smooth ball of mass 500 g is moving in the $xy$-plane and collides with a smooth fixed vertical wall which contains the line $x + y = 3$. The velocity of the ball just before the impact is $(-4\mathbf{i} - 2\mathbf{j})\,\text{ms}^{-1}$. The coefficient of restitution between the sphere and the wall is $\frac{1}{2}$. Find

  **a** the velocity of the ball immediately after the impact,

  **b** the kinetic energy lost as a result of the impact.

**4** A small smooth sphere of mass $m$ is moving with velocity $(6\mathbf{i} + 3\mathbf{j})\,\text{m s}^{-1}$ when it hits a smooth wall. It rebounds from the wall with velocity $(2\mathbf{i} - 2\mathbf{j})\,\text{m s}^{-1}$. Find

  **a** the magnitude and direction of the impulse received by the sphere,

  **b** the coefficient of restitution between the sphere and the wall.

**5** Two small smooth spheres $A$ and $B$ have equal radii. The mass of $A$ is $4m\,\text{kg}$ and the mass of $B$ is $m\,\text{kg}$. The spheres are moving on a smooth horizontal plane and they collide. Immediately before the collision the velocity of $A$ is $(2\mathbf{i} + 3\mathbf{j})\,\text{m s}^{-1}$ and the velocity of $B$ is $(3\mathbf{i} - \mathbf{j})\,\text{m s}^{-1}$. Immediately after the collision the velocity of $A$ is $(3\mathbf{i} + 2\mathbf{j})\,\text{m s}^{-1}$. Find

  **a** the velocity of $B$ immediately after the collision,

  **b** a unit vector parallel to the line of centres of the spheres at the instant of the collision.

**6**

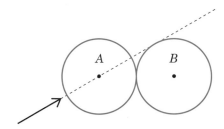

A smooth uniform sphere $B$ is at rest on a smooth horizontal plane, when it is struck by an identical sphere $A$ moving on the plane. Immediately before the impact, the line of motion of the centre of $A$ is tangential to the sphere $B$, as shown in the diagram above. The coefficient of restitution between the spheres is $\frac{2}{3}$. The direction of motion of $A$ is turned through an angle $\theta$ by the impact.

Show that $\theta = \tan^{-1}\dfrac{5\sqrt{3}}{9}$.

**7**

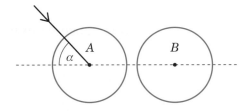

A smooth uniform sphere $A$, moving on a smooth horizontal table, collides with a second identical sphere $B$ which is at rest on the table. When the spheres collide the line joining their centres makes an angle of $\alpha$ with the direction of motion of $A$, as shown in the diagram. The direction of motion of $A$ is deflected through an angle $\theta$ by the collision. Given that $\alpha = \tan^{-1}\frac{3}{4}$ and that the coefficient of restitution between the spheres is $e$,

show that $\tan\theta = \dfrac{6 + 6e}{17 - 8e}$.

**8**

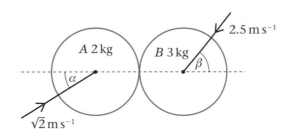

Two smooth uniform spheres $A$ and $B$ of equal radius have masses 2 kg and 3 kg respectively. They are moving on a smooth horizontal plane when they collide. Immediately before the collision the speed of $A$ is $\sqrt{2}$ m s$^{-1}$ and the speed of $B$ is 2.5 m s$^{-1}$. When they collide the line joining their centres makes an angle $\alpha$ with the direction of motion of $A$ and an angle $\beta$ with the direction of motion of $B$, where $\tan \alpha = 1$ and $\tan \beta = \frac{3}{4}$, as shown in the diagram. The coefficient of restitution between $A$ and $B$ is $\frac{2}{3}$.

Find the speed of each sphere after the collision.

**9**  A red ball is stationary on a rectangular billiard table $OABC$. It is then struck by a white ball of equal mass and equal radius moving with velocity $u \, (-2\mathbf{i} + 8\mathbf{j})$ where $\mathbf{i}$ and $\mathbf{j}$ are unit vectors parallel to $OA$ and $OC$ respectively. After the impact the velocity of the red ball is parallel to the vector $(-\mathbf{i} + \mathbf{j})$ and the velocity of the white ball is parallel to the vector $(2\mathbf{i} + 4\mathbf{j})$. Prove that the coefficient of restitution between the two balls is $\frac{3}{5}$.

**10**

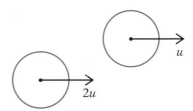

Two uniform spheres, each of mass $m$ and radius $a$, collide when moving on a horizontal plane. Before the impact the spheres are moving with speeds $2u$ and $u$, as shown in the diagram.

The centres of the spheres are moving on parallel paths distance $\dfrac{6a}{5}$ apart.

The coefficient of restitution between the spheres is $\frac{3}{4}$. Find the speeds of the spheres just after the impact, and show that the angle between their paths is then equal to $\tan^{-1} \frac{14}{23}$.

# Summary of key points

1  In this chapter we have modelled the moving objects as particles, and the surfaces as smooth. Our models do not take any account of the possibility of any friction or that objects could be rolling or spinning. The answers we have found give an approximation to what actually happens.

2  In an impact between a smooth sphere and a fixed surface:
   - The impulse on the sphere acts perpendicular to the surface, through the centre of the sphere.
   - Newton's law of restitution applies to the component of the velocity of the sphere perpendicular to the surface.
   - The component of the velocity of the sphere parallel to the surface is unchanged.

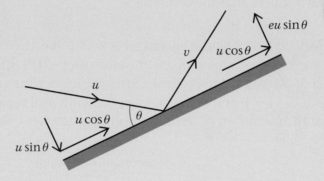

3  In an impact between two spheres:

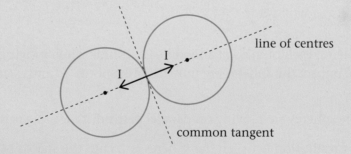

   - The reaction between the two spheres acts along the line of centres, so the impulse affecting each sphere also acts along the line of centres.
   - The components of the velocities of the spheres perpendicular to the line of centres are unchanged in the impact.
   - Newton's law of restitution applies to the components of the velocities of the spheres parallel to the line of centres.
   - The law of conservation of momentum applies parallel to the line of centres.

After completing this chapter you should be able to:
- solve problems involving motion in a straight line where the resistance to motion varies with speed
- model an object moving in a straight line under gravity when the resistance to motion varies with speed
- solve problems involving the motion of a vehicle with an engine working at a constant rate.

# Resisted motion of a particle moving in a straight line

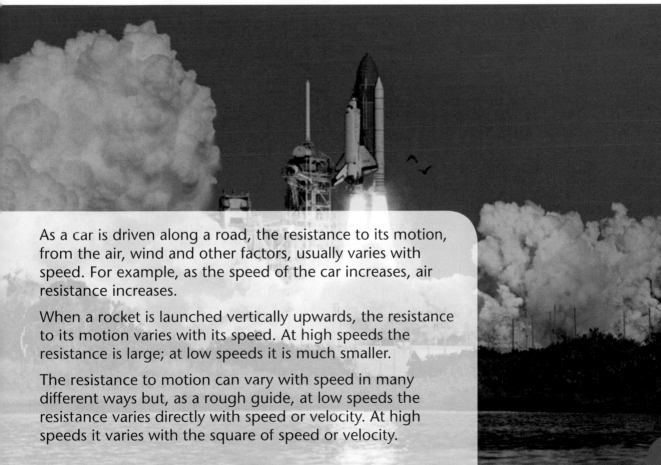

As a car is driven along a road, the resistance to its motion, from the air, wind and other factors, usually varies with speed. For example, as the speed of the car increases, air resistance increases.

When a rocket is launched vertically upwards, the resistance to its motion varies with its speed. At high speeds the resistance is large; at low speeds it is much smaller.

The resistance to motion can vary with speed in many different ways but, as a rough guide, at low speeds the resistance varies directly with speed or velocity. At high speeds it varies with the square of speed or velocity.

## 3.1 You can use calculus when a particle moves in a straight line against a resistance which varies with the speed of the particle.

The displacement from a fixed point ($x$), velocity ($v$) and acceleration ($a$) when the acceleration of a particle is varying with time are connected by the relations

$$a = \frac{dv}{dt} = \frac{d^2x}{dt^2} \qquad ①$$

Using the chain rule for differentiation

$$a = \frac{dv}{dt} = \frac{dv}{dx} \times \frac{dx}{dt}$$

As $v = \dfrac{dx}{dt}$, then

$$a = \frac{dv}{dx} \times v = v\frac{dv}{dx} \qquad ②$$

Also, if you differentiate $\frac{1}{2}v^2$ implicitly with respect to $x$, you obtain

$$\frac{d}{dx}\left(\frac{1}{2}v^2\right) = \frac{1}{2} \times 2v \times \frac{dv}{dx} = v\frac{dv}{dx} \qquad ③$$

Combining the results ② and ③, you obtain $a = v\dfrac{dv}{dx} = \dfrac{d}{dx}\left(\frac{1}{2}v^2\right)$.

The alternative forms for the acceleration can be summarised

$$\blacksquare \quad a = \frac{dv}{dt} = \frac{d^2x}{dt^2} = v\frac{dv}{dx} = \frac{d}{dx}\left(\frac{1}{2}v^2\right)$$

With these alternative forms for the acceleration you can solve problems where the resistance to the motion of a particle varies with its speed or velocity. In this chapter, you will mainly use the forms $\dfrac{dv}{dt}$ and $v\dfrac{dv}{dx}$.

If the resistance is a function of velocity f($v$), then the equation of motion of the particle will often be a separable differential equation which you can solve using the method you learnt in book C4. In this book, the function f($v$) will always be in the form $a + bv$ or $a + bv^2$, where $a$ and $b$ are constants.

Displacements must be measured from a fixed point. You will often choose to measure displacements from the point from which a particle starts or the point from which it is projected.

■ **In forming an equation of motion, forces that tend to decrease the displacement are negative and forces that tend to increase the displacement are positive.**

## Example 1

A particle $P$ of mass $0.5\,\text{kg}$ moves in a straight horizontal line. When the speed of $P$ is $v\,\text{m s}^{-1}$, the resultant force acting on $P$ is a resistance of magnitude $3v\,\text{N}$.

Find the distance moved by $P$ as it slows down from $12\,\text{m s}^{-1}$ to $6\,\text{m s}^{-1}$.

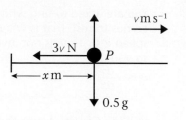

$R(\rightarrow)\ \mathbf{F} = m\mathbf{a}$

$-3v = 0.5a$

$-3v = 0.5v\dfrac{dv}{dx}$

$\dfrac{dv}{dx} = -6$

$v = \displaystyle\int -6\,dx$

$= -6x + A$

At $x = 0$, $v = 12$

$12 = -6 \times 0 + A \Rightarrow A = 12$

Hence

$v = 12 - 6x$

When $v = 6$

$6 = 12 - 6x$

$x = \dfrac{12 - 6}{6} = 1$

The distance moved by $P$ as it slows down from $12\,\text{m s}^{-1}$ to $6\,\text{m s}^{-1}$ is $1\,\text{m}$.

You measure the displacement of $P$, $x\,\text{m}$, from the point where it has a speed of $12\,\text{m s}^{-1}$.

As the resistance to motion is acting in the direction which decreases the displacement $x\,\text{m}$, the term, $3v$, representing the resistance in the equation of motion, has a negative sign.

When the question asks you to relate distance with speed, you choose the expression $a = v\dfrac{dv}{dx}$ for the acceleration.

The displacement is measured from the point where the speed of $P$ is $12\,\text{m s}^{-1}$. So you evaluate the constant of integration using $x = 0$ when $v = 12$.

## Example 2

A particle $P$ of mass $m$ is moving along $Ox$ in the direction of $x$ increasing. At time $t$, the only force acting on $P$ is a resistance of magnitude $mk(c^2 + v^2)$, where $v$ is the speed of $P$ and $c$ and $k$ are positive constants. When $t = 0$, $P$ is at $O$ and $v = U$. The particle $P$ comes to rest at the point $A$. Find

**a** the distance $OA$,

**b** the time $P$ takes to travel from $O$ to $A$.

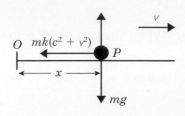

**a** $R(\rightarrow)$ $\qquad$ $F = ma$

$$-mk(c^2 + v^2) = ma$$

$$-\cancel{m}k(c^2 + v^2) = \cancel{m}v\frac{dv}{dx}$$

Separating the variables and integrating

$$\int \frac{v}{c^2 + v^2}\,dv = -\int k\,dx$$

$$\frac{1}{2}\ln(c^2 + v^2) = -kx + B$$

$$kx = B - \frac{1}{2}\ln(c^2 + v^2)$$

At $x = 0$, $v = U$

$$0 = B - \frac{1}{2}\ln(c^2 + U^2) \Rightarrow B = \frac{1}{2}\ln(c^2 + U^2)$$

Hence

$$kx = \frac{1}{2}\ln(c^2 + U^2) - \frac{1}{2}\ln(c^2 + v^2)$$

$$x = \frac{1}{2k}\ln\left(\frac{c^2 + U^2}{c^2 + v^2}\right)$$

At A, $v = 0$ and $x = OA$

$$OA = \frac{1}{2k}\ln\left(\frac{c^2 + U^2}{c^2}\right)$$

**b** $R(\rightarrow)$ $\qquad$ $F = ma$

$$-mk(c^2 + v^2) = ma$$

$$-\cancel{m}k(c^2 + v^2) = \cancel{m}\frac{dv}{dt}$$

Separating the variables and integrating

$$\int \frac{1}{c^2 + v^2}\,dv = -\int k\,dt$$

$$\frac{1}{c}\arctan\left(\frac{v}{c}\right) = -kt + D$$

Hence

$$kt = D - \frac{1}{c}\arctan\frac{v}{c}$$

When $t = 0$, $v = U$

$$0 = D - \frac{1}{c}\arctan\left(\frac{U}{c}\right) \Rightarrow D = \frac{1}{c}\arctan\left(\frac{U}{c}\right)$$

As the resistance to motion is acting in the direction which decreases the displacement $x$, the term, $mk(c^2 + v^2)$, representing the resistance in the equation of motion, has a negative sign.

You are asked to find where $P$ comes to rest, that is where its speed is 0. Time is not involved, so you use $a = v\dfrac{dv}{dx}$.

In book C4, you learnt that $\displaystyle\int \frac{f'(x)}{f(x)}\,dx = \ln f(x) + \text{a constant.}$ As $\dfrac{d}{dv}((c^2 + v^2)) = 2v$, then $\displaystyle\int \frac{v}{c^2 + v^2}\,dv = \frac{1}{2}\ln(c^2 + v^2).$ You need only put the arbitrary constant on one side of the equation.

You use the initial conditions to find the constant of integration.

Using the law of logarithms $\ln a - \ln b = \ln\left(\dfrac{a}{b}\right).$

The particle comes to rest where $v = 0$.

You are asked to find the time when $P$ comes to rest, that is where its speed is 0. Distance is not involved, so you use $a = \dfrac{dv}{dt}$.

The prerequisites given for this module require you to know that $\displaystyle\int \frac{1}{a^2 + x^2}\,dx = \frac{1}{a}\arctan\left(\frac{x}{a}\right).$ You need only put the arbitrary constant on one side of the equation.

Hence

$$kt = \frac{1}{c}\arctan\left(\frac{U}{c}\right) - \frac{1}{c}\arctan\left(\frac{V}{c}\right)$$

When $v = 0$

$$kt = \frac{1}{c}\arctan\left(\frac{U}{c}\right)$$ •——————— Using arctan $0 = 0$

$$t = \frac{1}{ck}\arctan\left(\frac{U}{c}\right)$$

The time $P$ takes to travel from $O$ to $A$

is $\frac{1}{ck}\arctan\left(\frac{U}{c}\right)$

## Example 3

A car of mass 800 kg travels along a straight horizontal road. The engine of the car produces a constant driving force of magnitude 2000 N. At time $t$ seconds, the speed of the car is $v\,\mathrm{m\,s^{-1}}$. As the car moves, the total resistance to the motion of the car is of magnitude $(400 + 4v^2)\,\mathrm{N}$. The car starts from rest.

**a** Find $v$ in terms of $t$.

**b** Show that the speed of the car cannot exceed $20\,\mathrm{m\,s^{-1}}$.

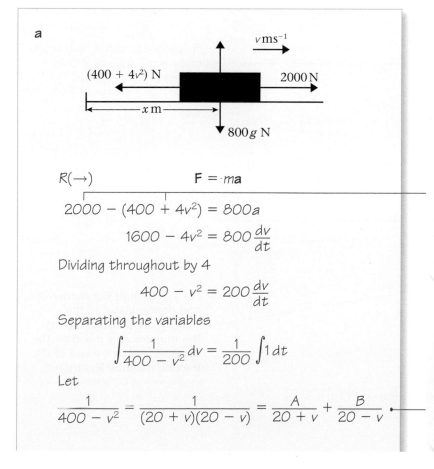

**a**

$R(\rightarrow)$ $\qquad$ $\mathbf{F} = m\mathbf{a}$

$$2000 - (400 + 4v^2) = 800a$$

$$1600 - 4v^2 = 800\frac{dv}{dt}$$

Dividing throughout by 4

$$400 - v^2 = 200\frac{dv}{dt}$$

Separating the variables

$$\int \frac{1}{400 - v^2}\,dv = \frac{1}{200}\int 1\,dt$$

Let

$$\frac{1}{400 - v^2} = \frac{1}{(20 + v)(20 - v)} = \frac{A}{20 + v} + \frac{B}{20 - v}$$ •

The driving force is in the direction of $x$ increasing and so the term representing the driving force, 2000, is positive in the equation of motion.

The resistance is in the direction of $x$ decreasing and so the term representing the resistance, $400 + 4v^2$, is negative in the equation of motion.

To integrate $\dfrac{1}{400 - v^2}$ you must factorise $400 - v^2 = (20 + v)(20 - v)$ and use partial fractions. To write $$\int \frac{1}{400 - v^2}\,dv = \ln(400 - v^2)$$ is a common error.

Multiplying throughout by $(20 + v)(20 - v)$

$$1 = A(20 - v) + B(20 + v)$$

Let $v = -20$

$$1 = 40A \Rightarrow A = \frac{1}{40}$$

Let $v = 20$

$$1 = 40B \Rightarrow B = \frac{1}{40}$$

Hence

$$\frac{1}{40} \int \left( \frac{1}{20 + v} + \frac{1}{20 - v} \right) dv = \frac{1}{200} \int 1\, dt$$

$$\frac{1}{40} \left( \ln(20 + v) - \ln(20 - v) \right) = \frac{1}{200} t + C$$

As $\ln a - \ln b = \ln\left(\frac{a}{b}\right)$ then
$\ln(20 + v) - \ln(20 - v) = \ln\left(\frac{20 + v}{20 - v}\right)$.

$$\ln\left(\frac{20 + v}{20 - v}\right) = \frac{1}{5} t + D, \text{ where } D = 40C$$

$$\frac{20 + v}{20 - v} = e^{\frac{t}{5} + D} = e^D e^{\frac{t}{5}} = F e^{\frac{t}{5}}$$

$e^D$ (e to an arbitrary constant) is another arbitrary constant $F$.

When $t = 0$, $v = 0$

$$\frac{20 + 0}{20 - 0} = F e^0 \Rightarrow F = 1$$

Hence

$$\frac{20 + v}{20 - v} = e^{\frac{t}{5}}$$

To complete part **a**, you must make $v$ the subject of this formula.

$$20 + v = 20\, e^{\frac{t}{5}} - v e^{\frac{t}{5}}$$

$$v(e^{\frac{t}{5}} + 1) = 20(e^{\frac{t}{5}} - 1)$$

$$v = \frac{20(e^{\frac{t}{5}} - 1)}{(e^{\frac{t}{5}} + 1)}$$

**b** For all real $t$, $e^{\frac{t}{5}} - 1 < e^{\frac{t}{5}} + 1$

As $-1 < 1$, for any real $x$, $x - 1 < x + 1$.

Hence

$$\frac{e^{\frac{t}{5}} - 1}{e^{\frac{t}{5}} + 1} < 1$$

and

$$\frac{20(e^{\frac{t}{5}} - 1)}{(e^{\frac{t}{5}} + 1)} < 20$$

For $t > 0$, both the numerator and denominator of this fraction are positive and, as the numerator is less than the denominator, the value of the fraction must be less than 1.

So the speed of the car cannot exceed $20\,\text{m s}^{-1}$, as required.

In Example 3, if a graph of $v$ against $t$ is plotted, you obtain

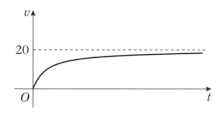

The speed approaches $20\,\mathrm{m\,s^{-1}}$ asymptotically. The speed approaches $20\,\mathrm{m\,s^{-1}}$ but cannot exceed it. Such a speed is called a **terminal** or **limiting** speed.

In this context, as the motion is in one direction, there is rarely any doubt as to the direction of motion and the **terminal** or **limiting** speed is often called, rather inaccurately, the **terminal** or **limiting** velocity.

■ As a body moves through a medium which resists its motion, its terminal or limiting speed can be found by assuming there is no resultant force acting on it and its acceleration is zero.

In Example 3, the equation of motion is

$$400 - v^2 = 200\frac{\mathrm{d}v}{\mathrm{d}t}.$$

If the acceleration $\left(= \dfrac{\mathrm{d}v}{\mathrm{d}t}\right)$ is 0, as $v$ is positive, this gives $v = 20$ and the terminal speed is $20\,\mathrm{m\,s^{-1}}$. This result is confirmed by the answers to parts **a** and **b**.

## Exercise 3A

**1** A particle $P$ of mass 2.5 kg moves in a straight horizontal line. When the speed of $P$ is $v\,\mathrm{m\,s^{-1}}$, the resultant force acting on $P$ is a resistance of magnitude $10v$ N. Find the time $P$ takes to slow down from $24\,\mathrm{m\,s^{-1}}$ to $6\,\mathrm{m\,s^{-1}}$.

**2** A particle $P$ of mass 0.8 kg is moving along the axis $Ox$ in the direction of $x$-increasing. When the speed of $P$ is $v\,\mathrm{m\,s^{-1}}$, the resultant force acting on $P$ is a resistance of magnitude $0.4v^2$ N. Initially $P$ is at $O$ and is moving with speed $12\,\mathrm{m\,s^{-1}}$. Find the distance $P$ moves before its speed is halved.

**3** A particle $P$ of mass 0.5 kg moves in a straight horizontal line against a resistance of magnitude $(4 + 0.5v)$ N, where $v\,\mathrm{m\,s^{-1}}$ is the speed of $P$ at time $t$ seconds. When $t = 0$, $P$ is at a point $A$ moving with speed $12\,\mathrm{m\,s^{-1}}$. The particle $P$ comes to rest at the point $B$. Find

**a** the time $P$ takes to move from $A$ to $B$,

**b** the distance $AB$.

**4** A particle of mass $m$ is projected along a rough horizontal plane with velocity $u\,\mathrm{m\,s^{-1}}$. The coefficient of friction between the particle and the plane is $\mu$. When the particle is moving with speed $v\,\mathrm{m\,s^{-1}}$, it is also subject to an air resistance of magnitude $kmgv^2$, where $k$ is a constant. Find the distance the particle moves before coming to rest.

**5** A particle $P$ of mass $m$ is moving along the axis $Ox$ in the direction of $x$-increasing. At time $t$ seconds, the velocity of $P$ is $v$. The only force acting on $P$ is a resistance of magnitude $k(a^2 + v^2)$. At time $t = 0$, $P$ is at $O$ and its speed is $U$. At time $t = T$, $v = \frac{1}{2}U$.

   **a** Show that $T = \dfrac{m}{ak}\left[\arctan\left(\dfrac{U}{a}\right) - \arctan\left(\dfrac{U}{2a}\right)\right]$.

   **b** Find the distance travelled by $P$ as its speed is reduced from $U$ to $\frac{1}{2}U$.

**6** A lorry of mass 2000 kg travels along a straight horizontal road. The engine of the lorry produces a constant driving force of magnitude 10 000 N. At time $t$ seconds, the speed of the lorry is $v\,\mathrm{m\,s^{-1}}$. As the lorry moves, the total resistance to the motion of the lorry is of magnitude $(4000 + 500v)$ N. The lorry starts from rest. Find

   **a** $v$ in terms of $t$,

   **b** the terminal speed of the lorry.

---

**3.2** **You can model the resistance to an object moving in a straight line under gravity using functions of velocity.**

---

- In book M1, you ignored air resistance when studying the motion of a particle projected vertically up or down. Air resistance, and other resistances, can be modelled using functions of velocity. Using these functions helps you to make more accurate predictions of a projectile's motion.

- Resistance always acts in the direction opposing motion. When a particle is moving vertically upwards, resistance acts downward. When a particle is moving vertically downward, resistance acts upwards. So when a particle is projected vertically upwards, its equation of motion as it ascends will be different from its equation of motion as it descends.

- In this book, you will only use functions of the form $a + bv$ or $a + bv^2$, where $a$ and $b$ are constants, to model resistance.

## Example **4**

A small ball of mass 0.5 kg is projected vertically upwards with velocity $20\,\mathrm{m\,s^{-1}}$ from a point $A$ on horizontal ground. Air resistance is modelled as a force of magnitude $0.2v\,$N, where $v\,\mathrm{m\,s^{-1}}$ is the velocity of the ball. Find the greatest height above $A$ attained by the ball.

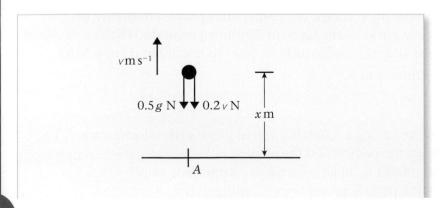

$R(\uparrow)$ $\qquad$ **F** $= m\mathbf{a}$

$$-0.5g - 0.2v = 0.5a$$

$$-4.9 - 0.2v = 0.5v\frac{dv}{dx}$$

> Both the weight of the ball and the resistance act in the direction of $x$ decreasing and so both are negative in the equation of motion.

Multiplying throughout by 10

$$-49 - 2v = 5v\frac{dv}{dx}$$

Separating the variables

$$\int 1\,dx = -5\int \frac{v}{49 + 2v}\,dv$$

> $\frac{v}{49 + 2v}$ is an improper fraction. The degree of the numerator is equal to the degree of the denominator. You must change the improper fraction into an expression with a proper fraction. You may use any method to do this.

Let

$$\frac{v}{49 + 2v} = A + \frac{B}{49 + 2v}$$

Multiplying throughout by $(49 + 2v)$

$$v = A(49 + 2v) + B$$

Equating coefficients of $v$

$$1 = 2A \Rightarrow A = \frac{1}{2}$$

Equating constant coefficients and using $A = \frac{1}{2}$

$$0 = 49A + B \Rightarrow B = -49A = -\frac{49}{2}$$

Hence

$$\int 1\,dx = -\frac{5}{2}\int \left(1 - \frac{49}{49 + 2v}\right)dv$$

$$x = -\frac{5v}{2} + \frac{5 \times 49}{4}\ln(49 + 2v) + C$$

> In book C4, you learnt that
> $$\int \frac{f'(x)}{f(x)}\,dx = \ln f(x) + \text{a constant.}$$
> As $\frac{d}{dv}((49 + 2v)) = 2$, then
> $$\int \frac{1}{49 + 2v}\,dv = \tfrac{1}{2}\ln(49 + 2v).$$
> You need only put the arbitrary constant on one side of the equation.

When $x = 0$, $v = 20$

$$0 = -50 + 61.25\ln 89 + C \Rightarrow C = 50 - 61.25\ln 89$$

Hence

$$x = 50 - \frac{5v}{2} + 61.25\ln(49 + v) - 61.25\ln 89$$

When $v = 0$

> The ball attains its greatest height when $v = 0$.

$$x = 50 + 61.25\ln 49 - 61.25\ln 89 \approx 13.4$$

The greatest height above A attained by the ball is 13.4 m (3 s.f.).

## Example 5

A particle $P$ of mass $m$ is released from rest at time $t = 0$ and falls vertically through a liquid. The motion $P$ is resisted by a force of magnitude $mkv^2$, where $v$ m s$^{-1}$ is the speed of $P$ at time $t$ seconds and $k$ is a positive constant. The terminal velocity of $P$ is $c$.

**a** Show that $c = \sqrt{\left(\dfrac{g}{k}\right)}$.

**b** Find $t$, in terms of $c$ and $g$, when $v = \frac{1}{2}c$.

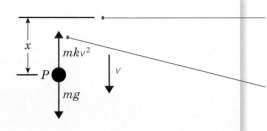

The displacement, $x$, must be measured from a fixed point. Here you can choose the point from which $P$ is released.

As $P$ is falling, the resistance opposes motion and acts upwards.

**a**   $R(\downarrow)$        $F = ma$

$\quad mg - mkv^2 = ma$

Dividing throughout by $m$

$\quad\quad g - kv^2 = a$       *

When $v = c$, $a = 0$

$\quad\quad g - kc^2 = 0 \Rightarrow c^2 = \dfrac{g}{k}$

$\quad\quad\quad c = \sqrt{\left(\dfrac{g}{k}\right)}$, as required

The weight acts in the direction of $x$ increasing, so, in this equation, the term $mg$ is positive. The resistance acts in the direction of $x$ decreasing, so, in this equation, the term $mkv^2$ is negative.

At the terminal speed, the acceleration is zero.

**b**   From part **a**

$\quad\quad c^2 = \dfrac{g}{k} \Rightarrow k = \dfrac{g}{c^2}$

Substituting for $k$ in * and using $a = \dfrac{dv}{dt}$

$\quad\quad g - \dfrac{g}{c^2}v^2 = \dfrac{dv}{dt}$

$\quad\quad \dfrac{g}{c^2}(c^2 - v^2) = \dfrac{dv}{dt}$

Separating the variables

$\quad\quad \displaystyle\int 1\, dt = \dfrac{c^2}{g}\int \dfrac{1}{c^2 - v^2}\, dv$

The question requires an answer in terms of $c$ and $g$ and not $k$, so you use the relation in part **a** between these variables to eliminate $k$ from the differential equation *.

To find $\displaystyle\int \dfrac{1}{c^2 - v^2}\, dv$, you must use partial fractions.

Let

$\quad\quad \dfrac{1}{c^2 - v^2} = \dfrac{1}{(c - v)(c + v)} = \dfrac{A}{c - v} + \dfrac{B}{c + v}$

Multiply throughout by $(c - v)(c + v)$

$\quad\quad 1 = A(c + v) + B(c - v)$

Let $v = c$

$$1 = 2cA \Rightarrow A = \frac{1}{2c}$$

Let $v = -c$

$$1 = 2cB \Rightarrow B = \frac{1}{2c}$$

Hence

$$\int 1\,dt = \frac{c^2}{g}\int \frac{1}{2c}\left(\frac{1}{c-v} + \frac{1}{c+v}\right)dv$$

$$t = \frac{c}{2g}\left[-\ln(c-v) + \ln(c+v)\right] + D$$

$$= \frac{c}{2g}\ln\left(\frac{c+v}{c-v}\right) + D$$

Using the log law

$\ln a - \ln b = \ln\left(\frac{a}{b}\right)$,

$\ln(c+v) - \ln(c-v) = \ln\left(\frac{c+v}{c-v}\right)$.

When $t = 0$, $v = 0$

$$0 = \frac{c}{2g}\ln 1 + D \Rightarrow D = 0$$

As $\ln 1 = 0$

Hence

$$t = \frac{c}{2g}\ln\frac{c+v}{c-v}$$

When $v = \frac{1}{2}c$

$$t = \frac{c}{2g}\ln\left(\frac{c+\frac{1}{2}c}{c-\frac{1}{2}c}\right) = \frac{c}{2g}\ln\left(\frac{\frac{3}{2}c}{\frac{1}{2}c}\right) = \frac{c}{2g}\ln 3$$

## Example 6

A particle $P$ of mass $m$ is projected vertically upwards with velocity $U$ from a point $A$ on horizontal ground. The particle $P$ is subject to air resistance of magnitude $mgkv^2$, where $v$ is the velocity of $P$ and $k$ is a positive constant. Find

**a** the greatest height above $A$ attained by $P$,

**b** the speed with which $P$ returns to $A$.

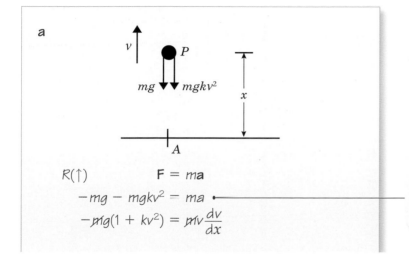

You measure the displacement, $x$, from the point from which $P$ is projected.

$R(\uparrow)$       $F = ma$

$$-mg - mgkv^2 = ma$$

$$-mg(1 + kv^2) = mv\frac{dv}{dx}$$

As $P$ ascends both the weight and the resistance act in the direction of $x$ decreasing and so both the terms $mg$ and $mgkv^2$ are negative in the equation of motion.

Separating the variables

$$\int 1\,dx = -\frac{1}{g}\int \frac{v}{1+kv^2}\,dv$$

$$x = A - \frac{1}{2gk}\ln(1+kv^2)$$

At $x = 0$, $v = U$

$$0 = A - \frac{1}{2gk}\ln(1+kU^2) \Rightarrow A = \frac{1}{2gk}\ln(1+kU^2)$$

$$x = \frac{1}{2gk}\ln(1+kU^2) - \frac{1}{2gk}\ln(1+kv^2)$$

$$= \frac{1}{2gk}\ln\left(\frac{1+kU^2}{1+kv^2}\right)$$

When $v = 0$

$$x = \frac{1}{2gk}\ln(1+kU^2)$$

The greatest height above A attained by P

is $\frac{1}{2gk}\ln(1+kU^2)$.

> $$\int \frac{2kv}{1+kv^2}\,dv = \ln(1+kv^2) + \text{constant,}$$
> and so
> $$\int \frac{v}{1+kv^2}\,dv = \frac{1}{2k}\ln(1+kv^2) + \text{constant.}$$

> At the greatest height, the speed of P is 0.

**b**

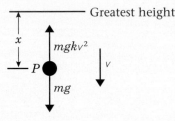

Greatest height

$$R(\downarrow) \qquad\qquad \mathbf{F} = m\mathbf{a}$$

$$mg - mgkv^2 = ma$$

$$\not{m}g(1-kv^2) = \not{m}v\frac{dv}{dx}$$

Separating the variables

$$\int 1\,dx = \frac{1}{g}\int \frac{v}{1-kv^2}\,dv$$

$$x = A - \frac{1}{2gk}\ln(1-kv^2)$$

At $x = 0$, $v = 0$

$$0 = A - \frac{1}{2gk}\ln(1) \Rightarrow A = 0$$

$$x = -\frac{1}{2gk}\ln(1-kv^2)$$

> In part **b**, you measure the displacement, $x$, of P from the point at which P attains its greatest height. In this part, as P moves downwards, the resistance acts upwards. That is in the opposite direction to the direction it acted in part **a**.

> As P descends, the weight acts in the direction of $x$ increasing and the resistance in the direction of $x$ decreasing. So the term $mg$ is positive and the term $mgkv^2$ is negative.

When $x = \dfrac{1}{2gk} \ln (1 + kU^2)$

$\dfrac{1}{2gk} \ln (1 + kU^2) = -\dfrac{1}{2gk} \ln (1 - kv^2)$

$\ln (1 + kU^2) = -\ln (1 - kv^2) = \ln \left( \dfrac{1}{1 - kv^2} \right)$

$1 + kU^2 = \dfrac{1}{1 - kv^2}$

$1 - kv^2 = \dfrac{1}{1 + kU^2}$

$kv^2 = 1 - \dfrac{1}{1 + kU^2} = \dfrac{1 + kU^2 - 1}{1 + kU^2} = \dfrac{kU^2}{1 + kU^2}$

$v^2 = \dfrac{U^2}{1 + kU^2} \Rightarrow v = \dfrac{U}{\sqrt{(1 + kU^2)}}$

$P$ returns to $A$ with speed $\dfrac{U}{\sqrt{(1 + kU^2)}}$.

> When $P$ reaches $A$, it has fallen a distance equal to the answer in part **a**.

> For any $x$, $-\ln x = \ln \left( \dfrac{1}{x} \right)$.

> Notice that if $k = 0$, that is if there is no air resistance, this becomes $v = U$, a result that is obvious from conservation of energy.

## Example 7

A particle of mass 2.5 kg moves under gravity down a line of greatest slope of a smooth plane inclined at an angle 30° to the horizontal. When the speed of the particle is $v$, air resistance to the motion of the particle has magnitude $0.4v$ N. The particle is released from rest at a point $A$. Four seconds after release, the particle passes through a point $B$. Find the speed of the particle at $B$.

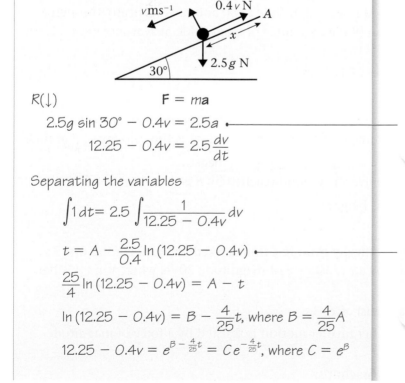

$R(\downarrow)$ $\qquad$ $\mathbf{F} = m\mathbf{a}$

$2.5g \sin 30° - 0.4v = 2.5a$

$12.25 - 0.4v = 2.5 \dfrac{dv}{dt}$

> The component of the weight, $2g \sin 30°$, acts down the plane, in the direction of $x$ increasing and so is positive in the equation of motion.

Separating the variables

$\displaystyle\int 1 \, dt = 2.5 \int \dfrac{1}{12.25 - 0.4v} \, dv$

$t = A - \dfrac{2.5}{0.4} \ln (12.25 - 0.4v)$

$\dfrac{25}{4} \ln (12.25 - 0.4v) = A - t$

$\ln (12.25 - 0.4v) = B - \dfrac{4}{25}t$, where $B = \dfrac{4}{25}A$

$12.25 - 0.4v = e^{B - \frac{4}{25}t} = Ce^{-\frac{4}{25}t}$, where $C = e^{B}$

> You are asked to find $v$ at a certain time, so make $v$ the subject of the formula. You can evaluate the arbitrary constant at any stage in your solution. In this question the constant has an easier form, which does not involve logarithms, if you evaluate it later in the solution.

Hence

$$0.4v = 12.25 - Ce^{-\frac{4}{25}t}$$

When $t = 0$, $v = 0$

$$0 = 12.25 - C \Rightarrow C = 12.25$$

$$0.4v = 12.25(1 - e^{-\frac{4}{25}t})$$

$$v = 30.625(1 - e^{-\frac{4}{25}t})$$

When $t = 4$

$$v = 30.625(1 - e^{-\frac{16}{25}}) \approx 14.5$$

The speed of the particle at $B$ is $14.5\,\text{m s}^{-1}$ (3 s.f.).

> As a numerical value of $g$ has been used, you should give your answer to 2 or 3 significant figures.

## Exercise 3B

**1** A particle $P$ of mass $0.8\,\text{kg}$ is projected vertically upwards with velocity $30\,\text{m s}^{-1}$ from a point $A$ on horizontal ground. Air resistance is modelled as a force of magnitude $0.02v^2\,\text{N}$, where $v\,\text{m s}^{-1}$ is the velocity of $P$.

Find the greatest height above $A$ attained by $P$.

**2** A particle $P$ of mass $1.5\,\text{kg}$ is released from rest at time $t = 0$ and falls vertically through a liquid. The liquid resists the motion of $P$ with a force of magnitude $5v\,\text{N}$ where $v\,\text{m s}^{-1}$ is the speed of $P$ at time $t$ seconds.

Find the value of $t$ when the speed of $P$ is $2\,\text{m s}^{-1}$.

**3** A small ball $B$ of mass $m$ is projected upwards from horizontal ground with speed $u$. Air resistance is modelled as a force of magnitude $mkv$, where $v\,\text{m s}^{-1}$ is the velocity of $P$ at time $t$ seconds.

**a** Show that the greatest height above the ground reached by $B$ is $\dfrac{u}{k} - \dfrac{g}{k^2}\ln\left(1 + \dfrac{ku}{g}\right)$.

**b** Find the time taken to reach this height.

**4** A parachutist of mass $60\,\text{kg}$ falls vertically from rest from a fixed balloon. For the first $3\,\text{s}$ of her motion, her fall is resisted by air resistance of magnitude $20v\,\text{N}$ where $v\,\text{m s}^{-1}$ is her velocity.

**a** Find the velocity of the parachutist after $3\,\text{s}$.

After $3\,\text{s}$, her parachute opens and her further motion is resisted by a force of magnitude $(20v + 60v^2)\,\text{N}$.

**b** Find the terminal speed of the parachutist.

**5** A particle $P$ of mass $m$ is projected vertically upwards with speed $u$ from a point $A$ on horizontal ground. The particle $P$ is subject to air resistance of magnitude $mgkv$, where $v$ is the speed of $P$ and $k$ is a positive constant.

**a** Find the greatest height above $A$ reached by $P$.

Assuming $P$ has not reached the ground,

**b** find an expression for the speed of the particle $t$ seconds after it has reached its greatest height.

**6** A particle of mass $m$ is projected vertically upwards from a point $A$ on horizontal ground with speed $u$. The particle reaches its greatest height above the ground at the point $B$.

**a** Ignoring air resistance, find the distance $AB$.

Instead of ignoring air resistance, it is modelled as a resisting force of magnitude $mkv^2$, where $v\,\text{m s}^{-1}$ is the velocity of the particle and $k$ is a positive constant. Using this model find

**b** the distance $AB$,

**c** the work done by air resistance against the motion of the particle as it moves from $A$ to $B$.

**7** A particle $P$ of mass $m$ is projected vertically downwards from a fixed point $O$ with speed $\dfrac{g}{2k}$, where $k$ is a constant. At time $t$ seconds after projection, the displacement of $P$ from $O$ is $x$ and the velocity of $P$ is $v$. The particle $P$ is subject to a resistance of magnitude $mkv$.

**a** Show that $v = \dfrac{g}{2k}(2 - e^{-kt})$.   **b** Find $x$ when $t - \dfrac{\ln 4}{k}$.

**8** A particle $P$ of mass $m$ is projected with speed $U$ up a rough plane inclined at an angle $30°$ to the horizontal. The coefficient of friction between $P$ and the plane is $\dfrac{\sqrt{3}}{4}$. The particle $P$ is subject to an air resistance of magnitude $mkv^2$, where $v$ is the speed of $P$ and $k$ is a positive constant.

Find the distance $P$ moves before coming to rest.

---

**3.3** You can use calculus when a particle, with an engine working at a constant rate, moves in a straight line against a resistance which varies with the speed of the particle.

In book M2, you studied work and power. You learnt that

$$\text{power} = \text{force} \times \text{velocity}.$$

If the engine of a vehicle is working at a constant rate $P$ and the engine generates a tractive force $F$ when the vehicle is travelling with velocity $v$ then

$$P = Fv$$

and hence

$$F = \frac{P}{v}.$$

So as the velocity of the vehicle increases, the tractive force generated by the engine decreases.

■ You can use the formula $F = \dfrac{P}{v}$ to form a separable differential equation when an engine working at a constant rate moves in a straight line against a resistance which varies with the speed of the particle.

## Example 8

A car of mass $800\,\text{kg}$ is moving along a straight horizontal road with the engine of the car working at $20\,\text{kW}$. The total resistance to the motion of the car is $25v\,\text{N}$, where $v\,\text{m s}^{-1}$ is the speed of the car at time $t$ seconds.

**a** Show that $32v\dfrac{\mathrm{d}v}{\mathrm{d}t} = 800 - v^2$.

**b** Find the time taken for the car to accelerate from $10\,\text{m s}^{-1}$ to $20\,\text{m s}^{-1}$.

**a**

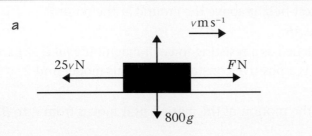

$20\,\text{kW} = 20\,000\,\text{W}$ •——— When forming an equation of motion you should use base SI units, so convert kW to W.

Let the tractive force generated by the engine be $F\,\text{N}$.

$$P = Fv$$

$$20\,000 = Fv \Rightarrow F = \frac{20\,000}{v}$$

$R(\rightarrow)$ $\qquad \mathbf{F} = m\mathbf{a}$

$$F - 25v = 800a$$

$$\frac{20\,000}{v} - 25v = 800\frac{\mathrm{d}v}{\mathrm{d}t} \quad\bullet$$

Divide this equation throughout by 25 and then put the left hand side over a common denominator.

$$\frac{800}{v} - v = 32\frac{\mathrm{d}v}{\mathrm{d}t}$$

$$\frac{800 - v^2}{v} = 32\frac{\mathrm{d}v}{\mathrm{d}t}$$

Hence

$$32v\frac{\mathrm{d}v}{\mathrm{d}t} = 800 - v^2, \text{ as required}$$

As $\dfrac{\mathrm{d}}{\mathrm{d}v}(800 - v^2) = -2v$, then

$\displaystyle\int \frac{-2v}{800 - v^2}\,\mathrm{d}v = \ln(800 - v^2) + C$ and

$\displaystyle\int \frac{v}{800 - v^2}\,\mathrm{d}v = B - \frac{1}{2}\ln(800 - v^2)$.

**b** Separating the variables

$$\int 1\,\mathrm{d}t = 32\int \frac{v}{800 - v^2}\,\mathrm{d}v$$

$$t = A - \frac{32}{2}\ln(800 - v^2)$$

Let $t = 0$, when $v = 10$ •——— You can measure time from the instant when the speed is $10\,\text{m s}^{-1}$.

$$0 = A - 16\ln(800 - 100) \Rightarrow A = 16\ln 700$$

Hence

$$t = 16\ln 700 - 16\ln(800 - v^2)$$

$$= 16\ln\left(\frac{700}{800 - v^2}\right)$$

When $v = 20$

$$t = 16 \ln\left(\frac{700}{800 - 400}\right) = 16 \ln\left(\frac{7}{4}\right) \approx 8.95$$

Either the exact or an approximate answer is acceptable.

The time taken to accelerate from $10 \, \text{m s}^{-1}$ to $20 \, \text{m s}^{-1}$ is $16 \ln\left(\frac{7}{4}\right) \text{s} = 8.95 \, \text{s}$ (3 s.f.)

## Example 9

A car of mass $m$ is moving along a horizontal road with its engine working at a constant rate $P$. There is a resistance to motion of the car of magnitude $mkv^2$, where $v$ is the speed of the car and $k$ is a constant. The terminal speed of the car is $W$.

**a** Show that $mk = \dfrac{P}{W^3}$

**b** Find, in terms of $k$, the distance travelled by the car as its speed increases from $\frac{1}{3}W$ to $\frac{2}{3}W$.

**a**

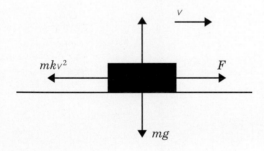

Let the tractive force generated by the engine be $F$.

$$P = Fv \Rightarrow F = \frac{P}{v}$$

$R(\rightarrow) \qquad \mathbf{F} = m\mathbf{a}$

$$\frac{P}{v} - mkv^2 = ma$$

Let $a = 0$, then $v = W$

You can find the terminal or limiting velocity by putting the acceleration equal to 0.

$$\frac{P}{W} - mkW^2 = 0 \Rightarrow P = mkW^3$$

Hence

$$mk = \frac{P}{W^3}, \text{ as required}$$

b The equation of motion in part **a** can be written as

$$\frac{P}{v} - mkv^2 = ma$$

$$\frac{P}{v} - \frac{P}{W^3}v^2 = mv\frac{dv}{dx}$$

$$\frac{P(W^3 - v^3)}{vW^3} = mv\frac{dv}{dx}$$

> You have to give the answer in terms of $k$ only and so you use the result of part **a** to eliminate $P$, $m$ and $k$. There are many alternative ways of doing this.

Separating the variables

$$\frac{P}{mW^3}\int 1\,dx = \int\frac{v^2}{W^3 - v^3}\,dv$$

As $k = \dfrac{P}{mW^3}$

$$k\int 1\,dx = \int\frac{v^2}{W^3 - v^3}\,dv$$

$$kx = A - \frac{1}{3}\ln(W^3 - v^3)$$

> An alternative approach is to use definite integrals, working out $k\displaystyle\int_0^x 1\,dx$ and $\displaystyle\int_{\frac{1}{3}W}^{\frac{2}{3}W}\frac{v^2}{W^3 - v^3}\,dv$.

Let $x = 0$, when $v = \frac{1}{3}W$

$$0 = A - \frac{1}{3}\ln\left(W^3 - \frac{W^3}{27}\right) \Rightarrow A = \frac{1}{3}\ln\left(\frac{26W^3}{27}\right)$$

Hence

$$kx = \frac{1}{3}\ln\left(\frac{26W^3}{27}\right) - \frac{1}{3}\ln(W^3 - v^3)$$

$$x = \frac{1}{3k}\ln\left[\frac{26W^3}{27(W^3 - v^3)}\right]$$

When $v = \frac{2}{3}W$

$$x = \frac{1}{3k}\ln\left[\frac{26W^3}{27(W^3 - \frac{8}{27}W^3)}\right] = \frac{1}{3k}\ln\left(\frac{26}{19}\right)$$

The distance travelled as the speed increases from $\frac{1}{3}W$ to $\frac{2}{3}W$ is $\frac{1}{3k}\ln\left(\frac{26}{19}\right)$.

## Exercise 3C

1 A car of mass 720 kg is moving along a straight horizontal road with the engine of the car working at 30 kW. At time $t = 0$, the car passes a point $A$ moving with speed 12 m s$^{-1}$. The total resistance to the motion of the car is $36v$ N, where $v$ m s$^{-1}$ is the speed of the car at time $t$ seconds.

Find the time the car takes to double its speed.

**2** A train of mass $m$ is moving along a straight horizontal track with its engine working at a constant rate of $16mkU^2$, where $k$ and $U$ are constants. The resistance to the motion of the train has magnitude $mkv$, where $v$ is the speed of the train.

Find the time the train takes to increase its speed from $U$ to $3U$.

**3** A van of mass 1200 kg is moving along a horizontal road with its engine working at a constant rate of 40 kW. The resistance to motion of the van is of magnitude of $5v^2$ N, where $v$ m s$^{-1}$ is the speed of the van. Find

**a** the terminal speed of the van,

**b** the distance the van travels while increasing its speed from 10 m s$^{-1}$ to 15 m s$^{-1}$.

**4** A car of mass $m$ is moving along a straight horizontal road with its engine working at a constant rate $D^3$. The resistance to the motion of the car is of magnitude $\dfrac{v^2}{k^3}$, where $v$ is the speed of the car and $k$ is a positive constant.

Find the distance travelled by the car as its speed increases from $\dfrac{kD}{4}$ to $\dfrac{kD}{2}$.

## Mixed exercise 3D

**1** A particle of mass $m$ moves in a straight line on a smooth horizontal plane in a medium which exerts a resistance of magnitude $mkv^2$, where $v$ is the speed of the particle and $k$ is a positive constant. At time $t = 0$ the particle has speed $U$.

Find, in terms of $k$ and $U$, the time at which the particle's speed is $\frac{3}{4}U$.   **E**

**2** A small pebble of mass $m$ is placed in a viscous liquid and sinks vertically from rest through the liquid. When the speed of the particle is $v$ the magnitude of the resistance due to the liquid is modelled as $mkv^2$, where $k$ is a positive constant.

Find the speed of the pebble after it has fallen a distance $D$ through the liquid.   **E**

**3** A car of mass 1000 kg is driven by an engine which generates a constant power of 12 kW. The only resistance to the car's motion is air resistance of magnitude $10v^2$ N, where $v$ m s$^{-1}$ is the speed of the car.

Find the distance travelled as the car's speed increases from 5 m s$^{-1}$ to 10 m s$^{-1}$.   **E**

**4** A bullet $B$, of mass $m$ kg, is fired vertically downwards into a block of wood $W$ which is fixed in the ground. The bullet enters $W$ with speed $U$ m s$^{-1}$ and $W$ offers a resistance of magnitude $m(14.8 + 5bv^2)$ N, where $v$ m s$^{-1}$ is the speed of $B$ and $b$ is a positive constant. The path of $B$ in $W$ remains vertical until $B$ comes to rest after travelling a distance $d$ metres into $W$.

Find $d$ in terms of $b$ and $U$.   **E**

**5** A particle of mass $m$ is projected vertically upwards, with speed $V$, in a medium which exerts a resisting force of magnitude $\dfrac{mgv^2}{c^2}$, where $v$ is the speed of the particle and $c$ is a positive constant.

**a** Show that the greatest height attained above the point of projection is $\dfrac{c^2}{2g}\ln\left(1 + \dfrac{V^2}{c^2}\right)$.

**b** Find an expression, in terms of $V$, $c$ and $g$, for the time to reach this height.   **E**

**6** A particle is projected vertically upwards with speed $U$ in a medium in which the resistance is proportional to the square of the speed. Given that $U$ is also the speed for which the resistance offered by the medium is equal to the weight of the particle show that

**a** the time of ascent is $\dfrac{\pi U}{4g}$,

**b** the distance ascended is $\dfrac{U^2}{2g}\ln 2$.

**7** At time $t$, a particle $P$, of mass $m$, moving in a straight line has speed $v$. The only force acting is a resistance of magnitude $mk(V_0^2 + 2v^2)$, where $k$ is a positive constant and $V_0$ is the speed of $P$ when $t = 0$.

**a** Show that, as $v$ reduces from $V_0$ to $\frac{1}{2}V_0$, $P$ travels a distance $\dfrac{\ln 2}{4k}$.

**b** Express the time $P$ takes to cover this distance in the form $\dfrac{\lambda}{kV_0}$, giving the value of $\lambda$ to two decimal places.

**8** A car of mass $m$ is moving along a straight horizontal road. When displacement of the car from a fixed point $O$ is $x$, its speed is $v$. The resistance to the motion of the car has magnitude $\dfrac{mkv^2}{3}$, where $k$ is a positive constant. The engine of the car is working at a constant rate $P$.

**a** Show that $3mv^2\dfrac{\mathrm{d}v}{\mathrm{d}x} = 3P - mkv^3$.

When $t = 0$, the speed of the car is half of its limiting speed.

**b** Find $x$ in terms of $m$, $k$, $P$ and $v$.

## Summary of key points

**1** The alternative forms of the acceleration are $a = \dfrac{\mathrm{d}v}{\mathrm{d}t} = \dfrac{\mathrm{d}^2x}{\mathrm{d}t^2} = v\dfrac{\mathrm{d}v}{\mathrm{d}x} = \dfrac{\mathrm{d}}{\mathrm{d}x}(\tfrac{1}{2}v^2)$.

**2** Displacements must be measured from a fixed point.

**3** In forming an equation of motion, forces that tend to decrease the displacement are negative and forces that tend to increase the displacement are positive.

**4** As a body moves through a medium which resists its motion, its **terminal** or **limiting** speed can be found by assuming there is no resultant force acting on it and its acceleration is zero.

**5** You can use the formula $F = \dfrac{P}{v}$ to form a separable differential equation when an engine working at a constant rate moves in a straight line against a resistance which varies with the speed of the particle.

# Review Exercise

**1** A river of width 40 m flows with uniform and constant speed between straight banks. A swimmer crosses as quickly as possible and takes 30 s to reach the other side. She is carried 25 m downstream.

Find

**a** the speed of the river,

**b** the speed of the swimmer relative to the water. *E*

**2** At noon, a boat $P$ is on a bearing of 120° from boat $Q$. Boat $P$ is moving due east at a constant speed of 12 km h$^{-1}$. Boat $Q$ is moving in a straight line with a constant speed of 15 km h$^{-1}$ on a course to intercept $P$. Find the direction of motion of $Q$, giving your answer as a bearing. *E*

**3** Points $A$ and $B$ are directly opposite each other on the parallel banks of a river. A motorboat, which travels at 4 m s$^{-1}$ relative to the water, crosses from $A$ to $B$. Given that the distance $AB$ is 400 m and that the river is flowing at 1.5 m s$^{-1}$ parallel to the banks, calculate

**a** the angle, to the nearest degree, between $AB$ and the direction in which the boat is being steered,

**b** the speed, in m s$^{-1}$ to 2 significant figures, of the motorboat relative to the bank,

**c** the time, to the nearest second, taken by the motorboat to cross the river. *E*

**4** A boy enters a large horizontal field and sees a friend 100 m due north. The friend is walking in an easterly direction at a constant speed of 0.75 m s$^{-1}$. The boy can walk at a maximum speed of 1 m s$^{-1}$.

Find the shortest time for the boy to intercept his friend and the bearing on which he must travel to achieve this. *E*

**5** A cyclist $P$ is cycling due north at a constant speed of 20 km h$^{-1}$. At 12 noon another cyclist $Q$ is due west of $P$. The speed of $Q$ is constant at 10 km h$^{-1}$.

Find the course which $Q$ should set in order to pass as close to $P$ as possible, giving your answer as a bearing. *E*

**6** [*In this question* **i** *and* **j** *are horizontal unit vectors due east and due north respectively*.]

An aeroplane makes a journey from a point $P$ to point $Q$ which is due east of $P$. The wind velocity is $w(\cos\theta\mathbf{i} + \sin\theta\mathbf{j})$, where $w$ is a positive constant. The velocity of the aeroplane relative to the wind is $v(\cos\phi\mathbf{i} - \sin\phi\mathbf{j})$, where $v$ is a constant and $v > w$. Given that $\theta$ and $\phi$ are both acute angles,

**a** show that $v\sin\phi = w\sin\theta$,

**b** find, in terms of $v$, $w$ and $\theta$, the speed of the aeroplane relative to the ground. **E**

**7** Boat $A$ is sailing due east at a constant speed of $10\,\text{km h}^{-1}$. To an observer on $A$, the wind appears to be blowing from due south. A second boat $B$ is sailing due north at a constant speed of $14\,\text{km h}^{-1}$. To an observer on $B$, the wind appears to be blowing from the south west. The velocity of the wind relative to the Earth is constant and is the same for both boats.

Find the velocity of the wind relative to the Earth, stating its magnitude and direction. **E**

**8** Ship $A$ is steaming on a bearing of $060°$ at $30\,\text{km h}^{-1}$ and at 9 a.m. it is $20\,\text{km}$ due west of a second ship $B$. Ship $B$ steams in a straight line.

**a** Find the least speed of $B$ if it is to intercept $A$.

Given that the speed of $B$ is $24\,\text{km h}^{-1}$,

**b** find the earliest time at which it can intercept $A$. **E**

**9** A cyclist $C$ is moving with a constant speed of $10\,\text{m s}^{-1}$ due south. Cyclist $D$ is moving with a constant speed of $16\,\text{m s}^{-1}$ on a bearing of $240°$.

**a** Show that the magnitude of the velocity of $C$ relative to $D$ is $14\,\text{m s}^{-1}$.

At 2 p.m., $D$ is $4\,\text{km}$ due east of $C$.

**b** Find

**i** the shortest distance between $C$ and $D$ during the subsequent motion,

**ii** the time, to the nearest minute, at which this shortest distance occurs. **E**

**10** A boat is sailing north at a speed of $15\,\text{km h}^{-1}$. To an observer on the boat the wind appears to blow from a direction $030°$.

The boat turns round and sails due south at the same speed. The velocity of the wind relative to the Earth remains constant, but to an observer on the boat it now appears to blow from $120°$.

Find the velocity of the wind relative to the Earth. **E**

**11** A pilot flying an aircraft at a constant speed of $2000\,\text{km h}^{-1}$ detects an enemy aircraft $100\,\text{km}$ away on a bearing of $045°$. The enemy aircraft is flying at a constant velocity of $1500\,\text{km h}^{-1}$ due west. Find

**a** the course, as a bearing to the nearest degree, that the pilot should set in order to intercept the enemy aircraft,

**b** the time, to the nearest s, that the pilot will take to reach the enemy aircraft. **E**

**12** At noon, two boats $A$ and $B$ are $6\,\text{km}$ apart with $A$ due east of $B$. Boat $B$ is moving due north at a constant speed of $13\,\text{km h}^{-1}$. Boat $A$ is moving with constant speed $12\,\text{km h}^{-1}$ and sets a course so as to pass as close as possible to boat $B$. Find

**a** the direction of motion of $A$, giving your answer as a bearing,

**b** the time when the boats are closest,

**c** the shortest distance between the boats. **E**

**13** A ship $A$ has maximum speed $30\,\mathrm{km\,h^{-1}}$. At time $t = 0$, $A$ is $70\,\mathrm{km}$ due west of $B$ which is moving at a constant speed of $36\,\mathrm{km\,h^{-1}}$ on a bearing of $300°$. Ship $A$ moves on a straight course at constant speed and intercepts $B$. The course of $A$ makes an angle $\theta$ with due north.

**a** Show that $-\arctan\frac{4}{3} \leqslant \theta \leqslant \arctan\frac{4}{3}$.

**b** Find the least time for $A$ to intercept $B$. **E**

**14** At 12 noon, ship $A$ is $20\,\mathrm{km}$ from ship $B$, on a bearing of $300°$. Ship $A$ is moving at a constant speed of $15\,\mathrm{km\,h^{-1}}$ on a bearing of $070°$. Ship $B$ moves in a straight line with constant speed $V\,\mathrm{km\,h^{-1}}$ and intercepts $A$.

**a** Find, giving your answer to 3 significant figures, the minimum possible value for $V$.

It is now given that $V = 13$.

**b** Explain why there are two possible times at which ship $A$ can intercept ship $B$.

**c** Find, giving your answer to the nearest minute, the earlier time at which ship $B$ can intercept ship $A$. **E**

**15** At time $t = 0$ particles $P$ and $Q$ start simultaneously from points which have position vectors $(\mathbf{i} - 2\mathbf{j} + 3\mathbf{k})\,\mathrm{m}$ and $(-\mathbf{i} + 2\mathbf{j} - \mathbf{k})\,\mathrm{m}$ respectively, relative to a fixed origin $O$. The velocities of $P$ and $Q$ are $(\mathbf{i} + 2\mathbf{j} - \mathbf{k})\,\mathrm{m\,s^{-1}}$ and $(2\mathbf{i} + \mathbf{k})\,\mathrm{m\,s^{-1}}$ respectively.

**a** Show that $P$ and $Q$ collide and find the position vector of the point at which they collide.

A third particle $R$ moves in such a way that its velocity relative to $P$ is parallel to the vector $(-5\mathbf{i} + 4\mathbf{j} - \mathbf{k})$ and its velocity relative to $Q$ is parallel to the vector $(-2\mathbf{i} + 2\mathbf{j} - \mathbf{k})$.

Given that all three particles collide simultaneously, find

**b i** the velocity of $R$,

**ii** the position vector of $R$ at time $t = 0$. **E**

**16** A rugby player is running due north with speed $4\,\mathrm{m\,s^{-1}}$. He throws the ball horizontally and the ball has an initial velocity relative to the player of $6\,\mathrm{m\,s^{-1}}$ in the direction $\theta°$ west of south, i.e. on a bearing of $(180 + \theta)°$, where $\tan\theta° = \frac{4}{3}$.

**a** Find the magnitude and direction of the initial velocity of the ball relative to a stationary spectator.

**b** Find also the bearing on which the ball appears to move initially to the referee who is running with speed $2\sqrt{2}\,\mathrm{m\,s^{-1}}$ in a north-westerly direction. **E**

**17** Two ships $A$ and $B$ are travelling with constant speeds $2u\,\mathrm{m\,s^{-1}}$ and $u\,\mathrm{m\,s^{-1}}$ respectively, $A$ on a bearing $\theta$ and $B$ on a bearing $90° + \theta$. It is also assumed that a third ship $C$ has a constant, but unknown, velocity which is taken to be a speed $v\,\mathrm{m\,s^{-1}}$ on a bearing $\phi$. To an observer on ship $B$ the velocity of $C$ appears to be due north.

**a** Show that $\dfrac{u}{\sin\phi} = \dfrac{v}{\cos\theta}$.

To an observer on ship $A$ the velocity of $C$ appears to be on a bearing of $135°$.

**b** Show that $2u(\cos\theta + \sin\theta) = v(\cos\phi + \sin\phi)$.

**c** Hence, find $\tan\phi$ in terms of $\tan\theta$.

Given that $\theta = 30°$ and $u = 10$,

**d** find the true velocity of $C$, giving your answer to 3 significant figures. **E**

**18** [*In this question* **i** *and* **j** *are horizontal unit vectors due east and due north respectively.*]

The airport $B$ is due north of airport $A$. On a particular day the velocity of the wind is $(70\mathbf{i} + 25\mathbf{j})$ km h$^{-1}$. Relative to the air, an aircraft flies with constant speed 250 km h$^{-1}$.

When the aircraft flies directly from $A$ to $B$, find

**a** its speed relative to the ground,

**b** its direction, as a bearing to the nearest degree, in which it must head.

After flying from $A$ to $B$, the aircraft returns directly to $A$.

**c** Calculate the ratio of the time taken on the outward journey to the time taken on the return flight. **E**

**19**

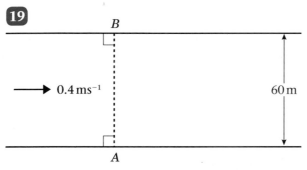

Mary swims in still water at 0.85 m s$^{-1}$. She swims across a straight river which is 60 m wide and flowing at 0.4 m s$^{-1}$. She sets off from a point $A$ on the near bank and lands at a point $B$, which is directly opposite $A$ on the far bank, as shown in the figure above.

Find

**a** the angle between the near bank and the direction in which Mary swims,

**b** the time she takes to cross the river.

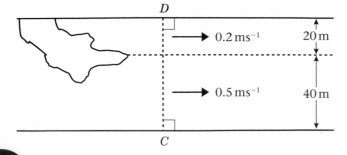

A little further downstream a large tree has fallen from the far bank into the river. The river is modelled as flowing at 0.5 m s$^{-1}$ for a width of 40 m from the near bank, and 0.2 m s$^{-1}$ beyond that. Nassim swims at 0.85 m s$^{-1}$ in still water. He swims across the river from a point $C$ on the near bank. The point $D$ on the far bank is directly opposite $C$ as shown above. Nassim swims at the same angle to the near bank as Mary.

**c** Find the maximum distance, downstream from $CD$, of Nassim during the crossing.

**d** Show that he will land at the point $D$. **E**

**20** A girl wishes to swim across a river from a fixed point $O$ on the bank, to a point $B$ on the opposite bank. The position vector of $B$ relative to $O$ is $20\mathbf{j}$ m. In a simple model the water is assumed to be flowing with uniform velocity $u\mathbf{i}$ m s$^{-1}$ and the girl intends to swim in such a way that she moves along the line $OB$.

**a** Given that $u = 0.6$ and that the speed of the girl relative to the water is 1 m s$^{-1}$, show that the time taken to swim across the river is 25 s.

A geographer points out that the flow of the river will be faster nearer the middle than closer to the banks and the model for the flow of the river is refined. When the girl is at a point $R$ on the river, with position vector $(x\mathbf{i} + y\mathbf{j})$ m, the velocity of the river at that point is $v\mathbf{i}$ m s$^{-1}$, where

$$v = \frac{y}{25}(20 - y), \quad 0 \leqslant y \leqslant 20.$$

The girl swims with velocity $(-p\mathbf{j} + q\mathbf{j})$ m s$^{-1}$ relative to the water, where $p$ and $q$ are positive constants. The girl starts to swim from $O$ at time $t = 0$ and the time taken to cross from $O$ to $B$ is now 50 s.

**b** Find the value of $q$ and hence show that, at time $t$ seconds, $y = 0.4t$.

**c** By considering the motion of the girl in the **i** direction, find the value of $p$.  **E**

**21** A smooth uniform sphere $S$ of mass $m$ is moving on a smooth horizontal plane when it collides with a fixed smooth vertical wall. Immediately before the collision, the speed of $S$ is $U$ and its direction of motion makes an angle $\alpha$ with the wall. The coefficient of restitution between $S$ and the wall is $e$.

Find the kinetic energy of $S$ immediately after the collision.  **E**

**22** A smooth sphere $S$, of mass $m$, is moving with speed $u$ on a horizontal plane when it collides with another smooth sphere, of mass $3m$ and having the same radius as $S$, which is at rest on the horizontal plane. The direction of motion of $S$ before impact makes an angle $\theta$, $0 < \theta < \frac{\pi}{2}$, with the line of centres of the two spheres. The coefficient of restitution between the spheres is $e$. After impact the spheres are moving in directions which are perpendicular to each other.

Find the value of $e$.  **E**

**23**

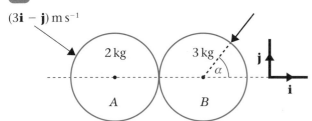

$(3\mathbf{i} - \mathbf{j})\,\mathrm{m\,s^{-1}}$

Two smooth uniform spheres $A$ and $B$, of equal radius, are moving on a smooth horizontal plane. Sphere $A$ has mass $2\,\mathrm{kg}$ and sphere $B$ has mass $3\,\mathrm{kg}$. The spheres collide and at the instant of collision the line joining their centres is parallel to **i**. Before the collision $A$ has velocity $(3\mathbf{i} - \mathbf{j})\,\mathrm{m\,s^{-1}}$ and after the collision it

has velocity $(-2\mathbf{i} - \mathbf{j})\,\mathrm{m\,s^{-1}}$. Before the collision the velocity of $B$ makes an angle $\alpha$ with the line of centres, as shown in the figure, where $\tan \alpha = 2$. The coefficient of restitution between the spheres is $\frac{1}{2}$.

Find, in terms of **i** and **j**, the velocity of $B$ before the collision.  **E**

**24** A small ball is moving on a horizontal plane when it strikes a smooth vertical wall. The coefficient of restitution between the ball and the wall is $e$. Immediately before the impact the direction of motion of the ball makes an angle of $60°$ with the wall. Immediately after the impact the direction of motion of the ball makes an angle of $30°$ with the wall.

**a** Find the fraction of the kinetic energy of the ball which is lost in the impact.

**b** Find the value of $e$.  **E**

**25** A smooth sphere $A$ moving with speed $u$ collides with an identical sphere $B$ which is at rest. The directions of motion of $A$ before and after impact makes angles $\frac{\pi}{3}$ and $\beta$ respectively with the line of centres at the moment of impact. The coefficient of restitution between the spheres is $0.8$.

Show that $\tan \beta = 10\sqrt{3}$.  **E**

**26** A smooth uniform sphere $P$ of mass $m$ is falling vertically and strikes a fixed smooth inclined plane with speed $u$. The plane is inclined at an angle $\theta$, $\theta < 45°$, to the horizontal.

The coefficient of restitution between $P$ and the inclined plane is $e$. Immediately after $P$ strikes the plane, $P$ moves horizontally.

**a** Show that $e = \tan^2 \theta$.

**b** Show that the magnitude of the impulse exerted by $P$ on the plane is $mu \sec \theta$.  **E**

**27**

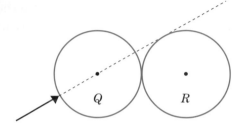

A smooth uniform sphere $R$ is at rest on a smooth horizontal plane, when it is struck by an identical sphere $Q$ moving on the plane. Immediately before the impact, the line of motion of the centre of $Q$ is tangential to the sphere $R$, as shown the figure. The direction of motion of $Q$ is turned through 30° by the impact.

Find the coefficient of restitution between the spheres.

**28**

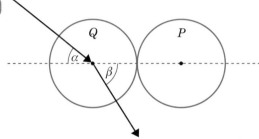

A smooth sphere $P$ lies at rest on a smooth horizontal plane. A second identical sphere $Q$, moving on the plane, collides with the sphere $P$. Immediately before the collision the direction of motion of $Q$ makes an angle $\alpha$ with the line joining the centres of the spheres. Immediately after the collision the direction of motion of $Q$ makes an angle $\beta$ with the line joining the centres of spheres, as shown in the figure. The coefficient of restitution between the spheres is $e$.

Show that $(1 - e)\tan\beta = 2\tan\alpha$.

**29** A smooth uniform sphere $S$ of mass $m$ is moving on a smooth horizontal table. The sphere $S$ collides with another smooth uniform sphere $T$, of the same radius as

$S$ but of mass $km$, $k > 1$, which is at rest on the table. The coefficient of restitution between the spheres is $e$. Immediately before the spheres collide the direction of motion of $S$ makes an angle $\theta$ with the line joining their centres. as shown in the figure.

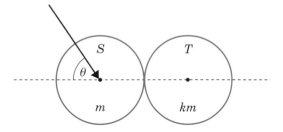

Immediately after the collision the directions of motion of $S$ and $T$ are perpendicular.

**a** Show that $e = \frac{1}{k}$.

Given that $k = 2$ and that the kinetic energy lost in the collision is one quarter of the initial kinetic energy,

**b** find the value of $\theta$.

**30** A smooth uniform sphere $A$ has mass $2m$ kg and another smooth uniform sphere $B$, with the same radius as $A$, has mass $m$ kg. The spheres are moving on a smooth horizontal plane when they collide. At the instant of collision the line joining the centres of the spheres is parallel to $\mathbf{j}$. Immediately **after** the collision, the velocity of $A$ is $(3\mathbf{i} - \mathbf{j})$ m s$^{-1}$ and the velocity of $B$ is $(2\mathbf{i} + \mathbf{j})$ m s$^{-1}$. The coefficient of restitution between the spheres is $\frac{1}{2}$.

**a** Find the velocities of the two spheres immediately before the collision.

**b** Find the magnitude of the impulse in the collision.

**c** Find, to the nearest degree, the angle through which the direction of motion of $A$ is deflected by the collision. **E**

**31**

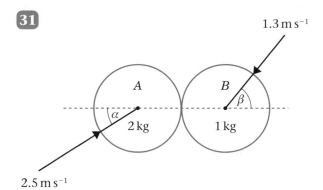

Two smooth uniform spheres $A$ and $B$ of equal radius have masses 2 kg and 1 kg respectively. They are moving on a smooth horizontal plane when they collide. Immediately before the collision the speed of $A$ is 2.5 m s$^{-1}$ and the speed of $B$ is 1.3 m s$^{-1}$. When they collide the line joining their centres makes an angle $\alpha$ with the direction of motion of $A$ and an angle $\beta$ with the direction of motion of $B$, where $\tan \alpha - \frac{4}{3}$ and $\beta = \frac{12}{5}$, as shown in the figure.

**a** Find the components of the velocities of $A$ and $B$ perpendicular and parallel to the line of centres immediately before the collision.

The coefficient of restitution between $A$ and $B$ is $\frac{1}{2}$.

**b** Find, to one decimal place, the speed of each sphere after the collision. **E**

**32**

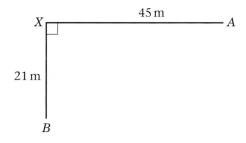

The figure represents the scene of a road accident. A car of mass 600 kg collided at the point $X$ with a stationary van of mass 800 kg. After the collision the van came to rest at the point $A$ having travelled a horizontal distance of 45 m, and the

car came to rest at the point $B$ having travelled a horizontal distance of 21 m. The angle $AXB$ is 90°.

The accident investigators are trying to establish the speed of the car before the collision and they model both vehicles as small spheres.

**a** Find the coefficient of restitution between the car and the van.

The investigators assume that after the collision, and until the vehicles came to rest, the van was subject to a constant horizontal force of 500 N acting along $AX$ and the car to a constant horizontal force of 300 N along $BX$.

**b** Find the speed of the car immediately before the collision. **E**

**33** A smooth sphere $T$ is at rest on a smooth horizontal table. An identical sphere $S$ moving on the table with speed $U$ collides with $T$. The directions of motion of $S$ before and after impact make angles of 30° and $\beta$° ($0 < \beta < 90$) respectively with $L$, the line of centres at the moment of impact. The coefficient of restitution between $S$ and $T$ is $e$.

**a** Show that $V$, the speed of $T$ immediately after impact, is given by
$$V = \frac{U\sqrt{3}}{4}(1 + e)$$

**b** Find the components of the velocity of $S$, parallel and perpendicular to $L$, immediately after impact.

Given that $e = \frac{2}{3}$,

**c** find, to 1 decimal place, the value of $\beta$. **E**

**34** Two small spheres $A$ and $B$, of equal size and of mass $m$ and $2m$ respectively, are moving initially with the same speed $U$ on a smooth horizontal floor. The spheres collide when their centres are on a line $L$. Before the collision the spheres are moving towards each other, with their directions of motion perpendicular to each other and each inclined at an angle 45° to the line $L$, as shown in the figure below. The coefficient of restitution between the spheres is $\frac{1}{2}$.

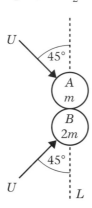

**a** Find the magnitude of the impulse which acts on $A$ in the collision.

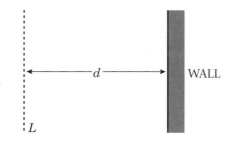

The line $L$ is parallel to and a distance $d$ from a smooth vertical wall, as shown in the second figure.

**b** Find, in terms of $d$, the distance between the points at which the spheres first strike the wall. **E**

**35** Two smooth uniform spheres $A$ and $B$ have equal radii. Sphere $A$ has mass $m$ and sphere $B$ has mass $km$. The spheres are at rest on a smooth horizontal table. Sphere $A$ is then projected along the table with speed $u$ and collides with $B$. Immediately before the collision, the direction of motion of $A$ makes an angle of 60° with the line joining the centres of the two spheres. The coefficient of restitution between the spheres is $\frac{1}{2}$.

**a** Show that the speed of $B$ immediately after the collision is $\dfrac{3u}{4(k+1)}$.

Immediately after the collision the direction of motion of $A$ makes an angle $\arctan(2\sqrt{3})$ with the direction of motion of $B$.

**b** Show that $k = \frac{1}{2}$.

**c** Find the loss of kinetic energy due to the collision. **E**

**36** Two equal smooth spheres approach each other from opposite directions with equal speeds. The coefficient of restitution between the spheres is $e$. At the moment of impact, their common normal is inclined at an angle $\theta$ to the original direction of motion. After impact, each sphere moves at right angles to its original direction of motion.

Show that $\tan\theta = \sqrt{e}$. **E**

**37**

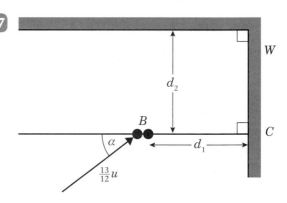

A small ball $Q$ of mass $2m$ is at rest at the point $B$ on a smooth horizontal plane. A second small ball $P$ of mass $m$ is moving on the plane with speed $\frac{13}{12}u$ and collides with $Q$. Both the balls are smooth, uniform and of the same radius. The point $C$ is on a smooth vertical wall $W$ which is at a distance $d_1$ from $B$, and $BC$ is perpendicular to $W$. A second smooth vertical wall is perpendicular to $W$ and at a distance $d_2$ from $B$. Immediately before the collision occurs, the direction of motion of $P$ makes an angle $\alpha$ with $BC$, as shown in the figure, where $\tan\alpha = \frac{5}{12}$.

The line of centres of $P$ and $Q$ is parallel to $BC$. After the collision $Q$ moves towards $C$ with speed $\frac{3}{5}u$.

**a** Show that, after the collision, the velocity components of $P$ parallel and perpendicular to $CB$ are $\frac{1}{5}u$ and $\frac{5}{12}u$ respectively.

**b** Find the coefficient of restitution between $P$ and $Q$.

**c** Show that when $Q$ reaches $C$, $P$ is at a distance $\frac{4}{3}d_1$ from $W$.

For each collision between a ball and a wall the coefficient of restitution is $\frac{1}{2}$.

Given that the balls collide with each other again,

**d** show that the time between the two collisions of the balls is $\dfrac{15d_1}{u}$,

**e** find the ratio $d_1 : d_2$.  **E**

**38**

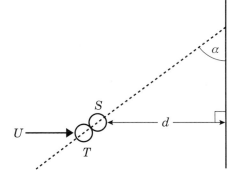

A small smooth uniform sphere $S$ is at rest on a smooth horizontal floor at a distance $d$ from a straight vertical wall. An identical sphere $T$ is projected along the floor with speed $U$ towards $S$ and in a direction which is perpendicular to the wall. At the instant when $T$ strikes $S$ the line joining their centres makes an angle $\alpha$ with the wall, as shown the figure.

Each sphere is modelled as having negligible diameter in comparison with $d$. The coefficient of restitution between the spheres is $e$.

**a** Show that the components of the velocity of $T$ after the impact, parallel

and perpendicular to the line of centres, are $\frac{1}{2}U(1 - e) \sin \alpha$ and $U \cos \alpha$ respectively.

**b** Show that the components of the velocity of $T$ after the impact, parallel and perpendicular to the wall are $\frac{1}{2}U(1 + e) \cos \alpha \sin \alpha$ and $\frac{1}{2}U[2 - (1 + e) \sin^2 \alpha]$ respectively.

The spheres $S$ and $T$ strike the wall at the points $A$ and $B$ respectively.

Given that $e = \frac{2}{3}$ and $\tan \alpha = \frac{3}{4}$,

**c** find, in terms of $d$, the distance $AB$.  **E**

**39**

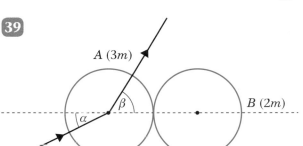

A uniform small smooth sphere of mass $3m$ moving with speed $u$ on a smooth horizontal table collides with a stationary small sphere $B$ of the same size as $A$ and of mass $2m$. The direction of motion of $A$ before impact makes an angle $\alpha$ with the line of centres of $A$ and $B$, and the direction of motion of $A$ after the impact makes an angle $\beta$ with the same line, as shown in the figure. The coefficient of restitution between the spheres is $\frac{2}{3}$.

**a** Show that $\tan \beta = 3 \tan \alpha$.

**b** Express $\tan (\beta - \alpha)$ in terms of $t$, where $t = \tan \alpha$.

**c** Hence find, as $\alpha$ varies, the maximum angle of deflection of $A$ caused by the impact.  **E**

**40** Two identical spheres $A$ and $B$ lie at rest on a smooth horizontal table. Sphere $B$ is projected along the table towards sphere $A$ with velocity $u\mathbf{i} + v\mathbf{j}$, where $\mathbf{i}$ is the unit vector along the line of centres at the time of impact and $\mathbf{j}$ is a unit vector perpendicular to $\mathbf{i}$ and in the plane of the table. Given that the coefficient of restitution between the spheres is $e$,

**a** find the velocities of the spheres after impact.

Given further that the velocity of $B$ before impact makes an angle $\theta$ with the direction of $\mathbf{i}$ and that the velocity of $B$ after impact makes an angle $\phi$ with the direction of $\mathbf{i}$,

**b** show that $\tan(\phi - \theta) = \dfrac{\tan\theta(1 + e)}{1 - e + 2\tan^2\theta}$.

**c** Hence show that, as $\theta$ varies, the maximum value of the angle of deviation, $\phi - \theta$, occurs when $\tan^2\theta = \dfrac{1 - e}{2}$. **E**

**41** A particle $P$ of mass $3\,\text{kg}$ moves in a straight line on a smooth horizontal plane. When the speed of $P$ is $v\,\text{m}\,\text{s}^{-1}$, the resultant force acting on $P$ is a resistance to motion of magnitude $2v\,\text{N}$.

Find the distance moved by $P$ while slowing down from $5\,\text{m}\,\text{s}^{-1}$ to $2\,\text{m}\,\text{s}^{-1}$. **E**

**42** A particle $P$ of mass $0.5\,\text{kg}$ is released from rest at time $t = 0$ and falls vertically through a liquid. The motion $P$ is resisted by a force of magnitude $2v\,\text{N}$, where $v\,\text{m}\,\text{s}^{-1}$ is the speed of $P$ at time $t$ seconds.

**a** Show that $5\dfrac{\mathrm{d}v}{\mathrm{d}t} = 49 - 20v$.

**b** Find the speed of $P$ when $t = 1$. **E**

**43** A particle of mass $m$ moves in a straight line on a horizontal table against a resistance of magnitude $\lambda(mv + k)$, where $\lambda$ and $k$ are constants. Given that the particle starts with speed $u$ at time $t = 0$, show that the speed $v$ of the particle at time $t$ is

$$v = \frac{k}{m}(e^{-\lambda t} - 1) + u e^{-\lambda t}.$$ **E**

**44** A particle $P$, of mass $m$, is projected upwards from horizontal ground with speed $U$. The motion takes place in a medium in which the resistance is of magnitude $\dfrac{mgv^2}{k^2}$, where $v$ is the speed of $P$ and $k$ is a positive constant.

Show that $P$ reaches its maximum height above ground after a time $T$ given by

$$T = \frac{k}{g}\arctan\left(\frac{U}{k}\right).$$ **E**

**45** At time $t = 0$ a particle of mass $m$ falls from rest at the point $A$ which is at a height $h$ above a horizontal plane. The particle is subject to a resistance of magnitude $mkv^2$, where $v$ is the speed of the particle at time $t$ and $k$ is a positive constant. The particle strikes the plane with speed $V$.

Show that $kV^2 = g(1 - e^{-2hk})$. **E**

**46** A particle of mass $m$ is projected vertically upwards with speed $U$. It is subject to air resistance of magnitude $mkv^2$, where $v$ is its speed and $k$ is a positive constant.

**a** Show that the greatest height of the particle above its point of projection is

$$\frac{1}{2k}\ln\left(1 + \frac{kU^2}{g}\right).$$

**b** Find an expression for the total work done against air resistance during the upward motion. **E**

**47** A lorry of mass $M$ is moving along a straight horizontal road. The engine produces a constant driving force of magnitude $F$. The total resistance to motion is modelled as having constant magnitude $kv^2$, where $k$ is a constant, and $v$ is the speed of the lorry.

Given that the lorry moves with constant speed $V$,

**a** show that $V = \sqrt{\left(\dfrac{F}{k}\right)}$.

Given instead that the lorry starts from rest,

**b** show that the distance travelled by the lorry in attaining a speed $\frac{1}{2}V$ is

$$\frac{M}{2k}\ln\left(\frac{4}{3}\right).$$ **E**

**48** A train of mass $m$ is moving along a straight horizontal railway line. At time $t$, the train is moving with speed $v$ and the resistance to motion has magnitude $kv$, where $k$ is a constant. The engine of the train is working at a constant rate $P$.

**a** Show that, when $v > 0$, $mv\dfrac{dv}{dt} + kv^2 = P$.

When $t = 0$, the speed of the train is $\frac{1}{3}\sqrt{\left(\dfrac{P}{k}\right)}$.

**b** Find, in terms of $m$ and $k$, the time taken for the train to double its initial speed. **E**

**49** The engine of a car of mass $800\,\text{kg}$ works at a constant rate of $32\,\text{kW}$. The car travels along a straight horizontal road and the resistance to motion of the car is proportional to the speed of the car. At time $t$ seconds, $t \geqslant 0$, the car has a speed $v\,\text{m\,s}^{-1}$ and when $t = 0$, its speed is $10\,\text{m\,s}^{-1}$.

**a** Show that

$$800v\frac{dv}{dt} = 32\,000 - kv^2,$$

where $k$ is a positive constant.

Given that the limiting speed of the car is $40\,\text{m\,s}^{-1}$, find

**b** the value of $k$,

**c** $v^2$ in terms of $t$. **E**

**50** A car of mass $M$ kg is driven by an engine working at a constant power $RU$ watts, where $R$ and $U$ are positive constants. When the speed of the car is $v\,\text{m\,s}^{-1}$, the resistance to motion is $\dfrac{Rv^2}{U^2}$ newtons.

**a** Show that the acceleration of the car, $a\,\text{m\,s}^{-2}$, when its speed is $v\,\text{m\,s}^{-1}$, is given by $R(U^3 - v^3) = MU^2va$.

**b** Hence show that the distance, in m, travelled by the car as it increases its speed from $u_1\,\text{m\,s}^{-1}$ to $u_2\,\text{m\,s}^{-1}$ $(u_1 < u_2 < U)$ is

$$\frac{MU^2}{3R}\ln\left(\frac{U^3 - u_1^3}{U^3 - u_2^3}\right).$$ **E**

**51** A car of mass $780\,\text{kg}$ is moving along a straight horizontal road with the engine of the car working at $21\,\text{kW}$. The total resistance to the motion of the car is $(20v + 100)\,\text{N}$, where $v\,\text{m\,s}^{-1}$ is the speed of the car at time $t$ seconds.

**a** Show that $39v\dfrac{dv}{dt} = (30 - v)(35 + v)$.

**b** Find an expression for the time taken for the car to accelerate from $15\,\text{m\,s}^{-1}$ to $V\,\text{m\,s}^{-1}$. **E**

**52** A railway truck of mass $3000\,\text{kg}$ moves along a straight, horizontal railway line. When its speed is $v\,\text{m\,s}^{-1}$, it experiences a total resistance to motion of $(2000 + 5v^2)\,\text{N}$. A cable is attached to the truck, and the tension in the cable exerts a constant tractive force of $6500\,\text{N}$ on the truck.

**a** Find the time taken for the truck to accelerate from rest to a speed of $20\,\text{m\,s}^{-1}$, giving your answer in seconds to 3 significant figures.

When the speed of the truck is $20\,\text{m\,s}^{-1}$, the cable breaks.

**b** Find the time taken after the cable breaks for the truck to come to rest, giving your answer in seconds to 3 significant figures. **E**

**53** A particle $P$ of mass $m$ moves in a medium which produces a resistance of magnitude $mkv$, where $v$ is the speed of $P$ and $k$ is a constant. The particle $P$ is projected vertically upwards in this medium with speed $\dfrac{g}{k}$.

**a** Show that $P$ comes instantaneously to rest after time $\dfrac{\ln 2}{k}$.

**b** Find, in terms of $k$ and $g$, the greatest height above the point of projection reached by $P$. **(E)**

**54** A particle of mass $m$ moves under gravity down a line of greatest slope of a smooth plane inclined at an angle $\alpha$ to the horizontal. When the speed of the particle is $v$ the resistance to motion of the particle is $mkv$, where $k$ is a positive constant.

**a** Show that the limiting speed $c$ of the particle is given by $c = \dfrac{g \sin \alpha}{k}$.

The particle starts from rest.

**b** Show that the time $T$ taken to reach the speed of $\frac{1}{2}c$ is given by $T = \dfrac{1}{k} \ln 2$.

**c** Find, in terms of $c$ and $k$, the distance travelled by the particle in attaining the speed of $\frac{1}{2}c$. **(E)**

**55** A particle of mass $m$ is falling vertically under gravity in a resisting medium. The particle is released from rest. The speed $v$ of the particle at a distance $x$ from rest is given by

$$v^2 = 2kg\left[1 - e^{-\frac{x}{k}}\right],$$

where $k$ is a positive constant.

**a** Show that the magnitude of the resistance is $\dfrac{mv^2}{2k}$.

The particle is projected upwards in the same medium with speed $\sqrt{(2kg)}$.

**b** Show that the maximum height reached by the particle above the point of projection is $k \ln 2$.

**c** Find the time taken to reach the maximum height above the point of projection. **(E)**

**56** A ship of mass $m$ is propelled in a straight line through the water by a propeller which develops a constant force of magnitude $F$. When the speed of the ship is $v$, the water causes a drag of magnitude $kv$, where $k$ is a constant, to act on the ship. The ship starts from rest at time $t = 0$.

**a** Show that the ship reaches half of its theoretical maximum speed of $\dfrac{F}{k}$ when $t = \dfrac{m \ln 2}{k}$.

When the ship is moving with speed $\dfrac{F}{2k}$, an emergency occurs and the captain reverses the engines so that the propeller force, which remains of magnitude $F$, acts backwards.

**b** Show that the ship covers a further distance

$$\frac{mF}{k^2}\left[\frac{1}{2} - \ln\left(\frac{3}{2}\right)\right]$$

on its original course, which may be assumed to remain unchanged, before being brought to rest. **(E)**

After completing this chapter you should be able to:
- investigate the motion of a particle which is moving under the influence of a restoring force proportional to the particle's displacement and a resistance which is proportional to its speed
- investigate the motion of a particle which is moving under the influence of the above two forces and is also forced to oscillate with a frequency other than its natural one.

# Damped and forced harmonic motion

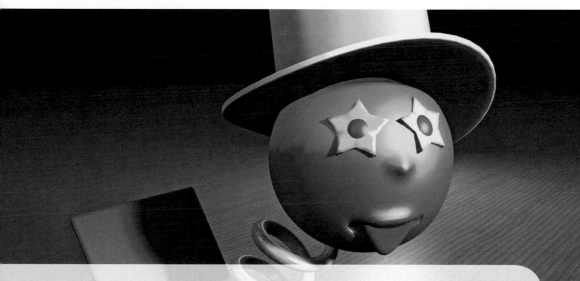

You studied simple harmonic motion (S.H.M.) in book M3, Chapter 3. When a particle is attached to an elastic string or spring and set in motion with no forces other than the tension/thrust and gravity acting on it, the motion is simple harmonic. The simple harmonic oscillations have constant amplitude. However, in practice, the amplitude of the oscillations decreases and the particle comes to rest fairly quickly. Hence the model must be refined to incorporate other factors such as air resistance and friction.

The toy in the picture can be held down and released. It will oscillate about its equilibrium position but the oscillations will not continue indefinitely. Soon after release the 'man' will come to rest.

**4.1** You can investigate the motion of a particle which is moving under the influence of a restoring force proportional to the particle's displacement and a resistance which is proportional to its speed.

■ For S.H.M. where the centre of oscillation is $O$ and the displacement of the particle from $O$ is $x$, the equation of motion reduces to

$$\ddot{x} = -\omega^2 x$$

■ When the model includes a resistance whose magnitude is proportional to the particle's speed, the resistance can be written as $mkv$. The equation of motion is now

$$m\frac{d^2x}{dt^2} = -m\omega^2 x - mkv$$

and writing $\frac{dx}{dt}$ instead of $v$ gives

$$\frac{d^2x}{dt^2} = -\omega^2 x - k\frac{dx}{dt}$$

or $\frac{d^2x}{dt^2} + k\frac{dx}{dt} + \omega^2 x = 0$

To investigate the motion of the particle, the solution of above equation must be obtained. The solution of this type of second order differential equation is studied in book FP2, Chapter 5, so you must study that chapter before proceeding further with these oscillations.

■ Oscillations of this type are called **damped harmonic motion**. There are three separate cases corresponding to the auxiliary equation having real distinct, equal or complex roots.

When $k^2 > 4\omega^2$ there two real distinct roots for the auxiliary equation. This is known as **heavy damping**. In this case there will be no oscillations performed as the resistive force is large compared with the restoring force.

When $k^2 = 4\omega^2$ the auxiliary equation has equal roots. This is known as **critical damping**. Again there will be no oscillations performed.

When $k^2 < 4\omega^2$ the auxiliary equation has complex roots. This is known as **light damping** and is the only case where oscillations are seen. The amplitude of the oscillations will decrease and the particle will come to rest.

For heavy and critical damping the exact nature of the motion will depend on the initial conditions given.

For light damping the period of the observed oscillations can be calculated. This is shown in Example 4.

■ The method for solution follows the same pattern in all cases.
  • Obtain the equation of motion for the particle. This will be a second order differential equation.
  • Solve the equation, using the methods of book FP2, Chapter 5.
  • Use the information in the question to obtain the arbitrary constants included in the general solution.

## Example 1

A particle $P$ of mass 0.5 kg moves in a horizontal straight line. At time $t$ seconds, the displacement of $P$ from a fixed point, $O$, of the line is $x$ metres and the speed of $P$ is $v \, \text{m s}^{-1}$. A force of magnitude $8x$ acts on $P$ in the direction $PO$. The particle is also subject to a resistance of magnitude $4v$. When $t = 0$, $x = 1.5$ and $P$ is moving in the direction of increasing $x$ with speed $4 \, \text{m s}^{-1}$.

**a** Show that $\dfrac{\text{d}^2 x}{\text{d}t^2} + 8\dfrac{\text{d}x}{\text{d}t} + 16x = 0$

**b** Find the value of $x$ when $t = 1$.

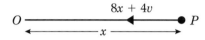

**a**
$$F = ma$$
$$-(8x + 4v) = 0.5\ddot{x}$$
$$0.5\ddot{x} + 4\dot{x} + 8x = 0$$
$$\frac{\text{d}^2 x}{\text{d}t^2} + 8\frac{\text{d}x}{\text{d}t} + 16x = 0$$

**b** Auxiliary equation:
$$m^2 + 8m + 16 = 0$$
$$(m + 4)^2 = 0$$
$$m = -4$$

Solve the differential equation using the methods of book FP2, Chapter 5.

General solution:
$$x = (\alpha + \beta t)e^{-4t}$$
$$t = 0, \, x = 1.5 \quad 1.5 = \alpha$$
$$v = \frac{\text{d}x}{\text{d}t} = \beta e^{-4t} - 4(\alpha + \beta t)\, e^{-4t}$$
$$t = 0, \, v = 4 \quad 4 = \beta - 4\alpha$$
$$\beta = 4 + 4\alpha = 4 + 4 \times 1.5 = 10$$

Use the initial conditions given in the question to obtain values for $\alpha$ and $\beta$.

$$\therefore x = (1.5 + 10t)\, e^{-4t}$$
$$t = 1 \quad x = 11.5\, e^{-4} = 0.2106\ldots$$

When $t = 1$, $x = 0.211$ (3 s.f.)

## Example 2

A particle $P$ of mass $m$ is hanging in equilibrium attached to one end of a light elastic spring of modulus of elasticity $mg$ and natural length $l$. The other end of the spring is attached to a fixed point. The particle is pulled vertically down a distance $\frac{1}{2}l$ from its equilibrium position and released from rest. At time $t$ the particle is at a distance $x$ from its equilibrium position and its speed is $v$. The air resistance opposing $P$'s motion has magnitude $m\sqrt{\frac{g}{l}}\,v$.

**a** Show that $\dfrac{d^2x}{dt^2} + \sqrt{\left(\dfrac{g}{l}\right)}\dfrac{dx}{dt} + \dfrac{g}{l}x = 0$.

**b** Find an expression for $x$ in terms of $t$.

**a**

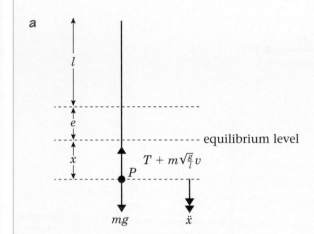

In equilibrium $\qquad T = \dfrac{\lambda x}{l} = \dfrac{mge}{l}$   •————————— Use Hooke's Law.

and $\qquad\qquad\quad T = mg$  •————————— R(↑)

$\therefore \qquad\qquad\quad \dfrac{mge}{l} = mg$

At distance $x$ below the equilibrium position:

$$F = ma$$

$$mg - T - m\sqrt{\frac{g}{l}}\,v = m\ddot{x}$$

Hooke's law: $T = \dfrac{mg(x + e)}{l}$

$$mg - \frac{mg(x + e)}{l} - m\sqrt{\frac{g}{l}}\frac{dx}{dt} = m\frac{d^2x}{dt^2}$$

$$-\frac{g}{l}x - \sqrt{\frac{g}{l}}\frac{dx}{dt} = \frac{d^2x}{dt^2}$$  •————————— From above, $\dfrac{mge}{l} = mg$

$$\frac{d^2x}{dt^2} + \sqrt{\frac{g}{l}}\frac{dx}{dt} + \frac{g}{l}x = 0$$

**b** Auxiliary equation: $m^2 + \sqrt{\dfrac{g}{l}}\, m + \dfrac{g}{l} m = 0$ —————— Solve the differential equation using the methods of book FP2, Chapter 5.

$$m = \frac{-\sqrt{\dfrac{g}{l}} \pm \sqrt{\dfrac{g}{l} - 4 \times \dfrac{g}{l}}}{2} = \frac{1}{2}\left(-\sqrt{\dfrac{g}{l}} \pm \sqrt{3}i\sqrt{\dfrac{g}{l}}\right) = \frac{1}{2}\left(\sqrt{\dfrac{g}{l}} -1 \pm \sqrt{3}i\right)$$

General solution: $x = e^{-\frac{1}{2}\sqrt{\frac{g}{l}}t}\left(A \cos \dfrac{\sqrt{3}}{2}\sqrt{\dfrac{g}{l}}t + B \sin \dfrac{\sqrt{3}}{2}\sqrt{\dfrac{g}{l}}t\right)$

When $t = 0$, $x = \dfrac{1}{2}l \;\pm\; \dfrac{1}{2}l = A$ •—————

Use the initial conditions given in the question to obtain values for $A$ and $B$.

$$\dot{x} = e^{-\frac{1}{2}\sqrt{\frac{g}{l}}t}\left(A \cos \dfrac{\sqrt{3}}{2}\sqrt{\dfrac{g}{l}}t + B \sin \dfrac{\sqrt{3}}{2}\sqrt{\dfrac{g}{l}}t\right) \times -\dfrac{1}{2}\sqrt{\dfrac{g}{l}}$$

$$+ e^{-\frac{1}{2}\sqrt{\frac{g}{l}}t}\left(-A \dfrac{\sqrt{3}}{2}\sqrt{\dfrac{g}{l}} \sin \dfrac{\sqrt{3}}{2}\sqrt{\dfrac{g}{l}}t + B\dfrac{\sqrt{3}}{2}\sqrt{\dfrac{g}{l}} \cos \dfrac{\sqrt{3}}{2}\sqrt{\dfrac{g}{l}}t\right)$$

When $t = 0$, $\dot{x} = 0 \therefore 0 = -\dfrac{1}{2}\sqrt{\dfrac{g}{l}}A + B\dfrac{\sqrt{3}}{2}\sqrt{\dfrac{g}{l}}$

$$0 = -\dfrac{1}{2}A + \dfrac{\sqrt{3}}{2}B$$

$$B = \dfrac{1}{\sqrt{3}}A = \dfrac{1}{2\sqrt{3}}l$$

$$\therefore x = e^{-\frac{1}{2}\sqrt{\frac{g}{l}}t}\left(\dfrac{1}{2}l \cos \dfrac{\sqrt{3}}{2}\sqrt{\dfrac{g}{l}}t + \dfrac{1}{2\sqrt{3}}l \sin \dfrac{\sqrt{3}}{2}\sqrt{\dfrac{g}{l}}t\right)$$

$$= \dfrac{1}{6}le^{-\frac{1}{2}\sqrt{\frac{g}{l}}t}\left(3 \cos \dfrac{\sqrt{3}}{2}\sqrt{\dfrac{g}{l}}t + \sqrt{3} \sin \dfrac{\sqrt{3}}{2}\sqrt{\dfrac{g}{l}}t\right)$$

## Example 3

A particle $P$ of mass $m$ hangs freely in equilibrium attached to one end of a light elastic string of natural length $l$ and modulus of elasticity $5mk^2l$. The other end of the string is attached to a fixed point $A$. The extension in the string is $e$.

**a** Find $e$ in terms of $g$ and $k$.

The particle is now held at rest in a container of liquid at a point which is a distance $2l$ vertically below $A$. At time $t = 0$, $P$ is released and at time $t$ the extension of the string is $(x + e)$. The liquid exerts a resistance to motion on $P$ of magnitude $6mkv$, where $v$ is the speed of $P$.

Show that while $P$ remains in the liquid and the string is taut

**b** $\dfrac{\mathrm{d}^2x}{\mathrm{d}t^2} + 6mk\dfrac{\mathrm{d}x}{\mathrm{d}t} + 5k^2x = 0.$

**c** Find an expression for $x$ in terms of $t$, $g$, $k$ and $l$.

**a**

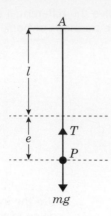

In equilibrium:

$T = \dfrac{\lambda x}{l} = \dfrac{5mk^2le}{l}$ •————————————— Use Hooke's Law.

$T = 5mk^2e$

$T = mg$ •————————————— R($\uparrow$)

$\therefore 5mk^2e = mg$

$e = \dfrac{g}{5k^2}$

**b**

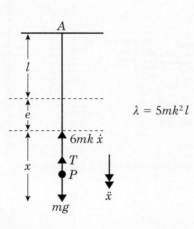

$\lambda = 5mk^2\,l$

$T = \dfrac{\lambda x}{l}$ •————————————— Use Hooke's Law.

$\quad = \dfrac{5mk^2l(x + e)}{l} = 5mk^2(x + e)$

$mg - T - 6mk\dot{x} = m\ddot{x}$ •————————————— Use $F = ma$.

$mg - 5mk^2(x + e) - 6mk\dot{x} = m\ddot{x}$

$\ddot{x} + 6k\dot{x} + 5k^2x = 0$ •————————————— From **a**, $5mk^2e = mg$.

or $\dfrac{d^2x}{dt^2} + 6k\dfrac{dx}{dt} + 4k^2x = 0$

**c** Auxiliary equation: $m^2 + 6km + 5k^2 = 0$

$$(m + 5)(m + 1) = 0$$

$$m = -5k \text{ or } -k$$

General solution: $x = Ae^{-5kt} + Be^{-kt}$

$t = 0 \quad x = l - \dfrac{g}{5k^2} \Rightarrow l - \dfrac{g}{5k^2} = A + B \quad \text{①}$

$\dot{x} = -5kAe^{-5kt} - kBe^{-kt}$

$t = 0 \quad \dot{x} = 0 \quad 0 = -5kA - kB$

$B = -5A \quad \text{②}$

In ① $\quad l - \dfrac{g}{5k^2} = A - 5A$

$4A = \dfrac{g}{5k^2} - l \quad A = \dfrac{g}{20k^2} - \dfrac{l}{4}$

$\therefore B = -5\left(\dfrac{g}{20k^2} - \dfrac{l}{4}\right)$

$\therefore x = \left(\dfrac{g}{20k^2} - \dfrac{l}{4}\right)e^{-5kt} - 5\left(\dfrac{g}{20k^2} - \dfrac{l}{4}\right)e^{-kt}$

or $x = \left(\dfrac{g}{20k^2} - \dfrac{l}{4}\right)(e^{-5kt} - 5e^{-kt})$

Solve the differential equation using the methods of book FP2, Chapter 5.

Use the initial conditions given in the question to obtain expressions for $A$ and $B$.

Example 4

One end of a light elastic spring of natural length $l$ and modulus of elasticity $2mk^2l$ is attached to a fixed point $A$. A particle $P$ of mass $m$ is attached to the other end and hangs in equilibrium vertically below $A$. The particle is pulled vertically downwards from its equilibrium position and released from rest. A resistance of magnitude $2mkv$, where $v$ is the speed of $P$, acts on $P$. At time $t$, the displacement of $P$ from its equilibrium position is $x$.

**a** Show that $\dfrac{d^2x}{dt^2} + 2k\dfrac{dx}{dt} + 2k^2x = 0$.

**b** Find the general solution of this differential equation.

**c** Write down the period of the oscillations.

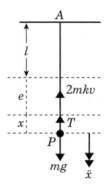

**a**   In equilibrium: $T = mg$  •————————————— R($\uparrow$)

$\qquad T = \dfrac{\lambda x}{l} = \dfrac{2mk^2le}{l} = 2mk^2e$ •

$\qquad \therefore 2mk^2e = mg$

At time t: $F = ma$

$mg - T - 2mkv = m\ddot{x}$

$T = \dfrac{2mk^2l(x + e)}{l} = 2mk^2(x + e)$ •

$\therefore mg - 2mk^2(x + e) - 2mkv = m\ddot{x}$

$-2mk^2x - 2mkv = m\ddot{x}$

$\ddot{x} + 2kv + 2k^2x = 0$

or $\dfrac{d^2x}{dt^2} + 2k\dfrac{dx}{dt} + 2k^2x = 0$

Use Hooke's law.

**b**   Auxiliary equation: $m^2 + 2km + 2k^2 = 0$ •

$\qquad m = \dfrac{-2k \pm \sqrt{4k^2 - 4 \times 2k^2}}{2} = k \pm ik$

$\qquad \therefore x = e^{-kt}(A\cos kt + B\sin kt)$

Solve the differential equation using the methods of book FP2, Chapter 5.

**c**   Period $= \dfrac{2\pi}{k}$ •

$(A\cos kt + B\sin kt)$ can be written as $R\cos(kt + \varepsilon)$ to give a period of $\dfrac{2\pi}{k}$.

Values are not needed for $R$ and $\varepsilon$.

## Exercise 4A

**1** A particle $P$ is moving in a straight line. At time $t$, the displacement of $P$ from a fixed point on the line is $x$. The motion of the particle is modelled by the differential equation

$$\frac{d^2x}{dt^2} + 4\frac{dx}{dt} + 8x = 0$$

When $t = 0$ $P$ is at rest at the point where $x = 2$.

**a** Find $x$ as a function of $t$.

**b** Calculate the value of $x$ when $t = \frac{\pi}{3}$.

**c** State whether the motion is heavily, critically or lightly damped.

**2** A particle $P$ is moving in a straight line. At time $t$, the displacement of $P$ from a fixed point on the line is $x$. The motion of the particle is modelled by the differential equation

$$\frac{d^2x}{dt^2} + 8\frac{dx}{dt} + 12x = 0$$

When $t = 0$ $P$ is at rest at the point where $x = 4$.

Find $x$ as a function of $t$.

**3** A particle $P$ is moving in a straight line. At time $t$, the displacement of $P$ from a fixed point on the line is $x$. The motion of the particle is modelled by the differential equation

$$\frac{d^2x}{dt^2} + 2\frac{dx}{dt} + 6x = 0$$

When $t = 0$ $P$ is at rest at the point where $x = 1$.

**a** Find $x$ as a function of $t$.

The smallest value of $t$, $t > 0$, for which $P$ is instantaneously at rest is $T$.

**b** Find the value of $T$.

**4** A particle $P$ of mass $m$ is attached to one end of a light elastic spring of natural length $l$ and modulus of elasticity $4m\omega^2 l$, where $\omega$ is a positive constant. The other end of the spring is attached to a fixed point $A$ and $P$ hangs in equilibrium vertically below $A$. At time $t = 0$, $P$ is projected vertically downwards with speed $u$. A resistance of magnitude $4m\omega v$, where $v$ is the speed of $P$, acts on $P$. The displacement of $P$ downwards from its equilibrium position at time $t$ is $x$.

**a** Show that $\dfrac{d^2x}{dt^2} + 4\omega\dfrac{dx}{dt} + 4\omega^2 x = 0$

**b** Find an expression for $x$ in terms of $u$, $t$ and $\omega$.

**c** Find the time at which $P$ comes to instantaneous rest.

**5** A particle $P$ of mass $m$ is attached to the mid-point of a light elastic string $AB$ of natural length $2a$ and modulus of elasticity $mg$. The ends $A$ and $B$ of the string are attached to fixed points on a smooth horizontal table with $AB = 4a$. The particle is released from rest at the point $C$ where $A$, $C$ and $B$ lie in a straight line and $AC = \frac{3}{2}a$. At time $t$ the displacement of $P$ from its equilibrium position is $x$. The particle is subject to a resisting force of magnitude $mnv$ where $v$ is the speed of $P$ and $n = \sqrt{\dfrac{2g}{a}}$.

**a** Show that $\dfrac{d^2x}{dt^2} + n\dfrac{dx}{dt} + n^2 x = 0$.

**b** Find an expression for $x$ in terms of $a$, $n$ and $t$.

**4.2** You can investigate the motion of a particle which is subject to the same two forces as in the previous section but is also forced to oscillate with a frequency other than its natural one.

■ **This type of motion is called forced harmonic motion.**

Suppose that, as in the previous section, a particle $P$ of mass $m$ kg is moving under the influence of two forces, a restoring force of magnitude $m\omega^2 x$, where $x$ is $P$'s displacement from a fixed point in the line of motion and a resistance of magnitude $mkv$, where $v$ is $P$'s speed. This time however, there is also another applied force of magnitude $m\mathrm{f}(t)$, which will vary with time.

■ **The equation of motion is now**

$$m\frac{d^2x}{dt^2} = -m\omega^2 x - mkv + m\mathrm{f}(t)$$

and writing $\dfrac{dx}{dt}$ instead of $v$ gives

$$\frac{d^2x}{dt^2} = -\omega^2 x - k\frac{dx}{dt} + \mathrm{f}(t)$$

or $\dfrac{d^2x}{dt^2} + k\dfrac{dx}{dt} + \omega^2 x = \mathrm{f}(t)$

Once again the equation of motion has reduced to a second order differential equation. The solution of this equation depends on the form of $\mathrm{f}(t)$. In book FP2, Chapter 5 you are shown that the solution is obtained by finding the general solution of the complementary equation $\dfrac{d^2x}{dt^2} + k\dfrac{dx}{dt} + \omega^2 x = 0$ and also a particular integral which takes account of $\mathrm{f}(t)$. (For full information about solving these equations refer to book FP2.)

■ **The method for solution follows the same pattern as for damped harmonic motion:**

- **Obtain the equation of motion for the particle. This will be a second order differential equation.**
- **Solve the equation, using the methods of book FP2, Chapter 5. Remember to obtain a particular integral as well as the complementary function.**
- **Use the information in the question to obtain the arbitrary constants included in the general solution.**

### Example 5

A particle $P$ of mass 1.5 kg is moving on the $x$-axis. At time $t$ the displacement of $P$ from the origin $O$ is $x$ metres and the speed of $P$ is $v$ m s$^{-1}$. Three forces act on $P$, namely a restoring force of magnitude $7.5x$ newtons, a resistance to the motion of $P$ of magnitude $6v$ newtons and a force of magnitude $12\sin t$ newtons acting in the direction $OP$. When $t = 0$, $x = 5$ and $\dfrac{dx}{dt} = 2$.

**a** Show that $\dfrac{d^2x}{dt^2} + 4\dfrac{dx}{dt} + 5x = 8\sin t$.

**b** Find $x$ as a function of $t$.

**c** Describe the motion when $t$ is large.

**a** $F = ma$

$$-7.5x - 6\frac{dx}{dt} + 12\sin t = 1.5\frac{d^2x}{dt^2}$$

$$\frac{d^2x}{dt^2} + 5x + 4\frac{dx}{dt} = 8\sin t$$

$$\frac{d^2x}{dt^2} + 4\frac{dx}{dt} + 5x = 8\sin t$$

**b** Auxiliary equation: $m^2 + 4m + 5 = 0$

> Solve the differential equation using the methods of book FP2, Chapter 5.

$$m = \frac{-4 \pm \sqrt{4^2 - 4 \times 5}}{2} = \frac{-4 \pm i\sqrt{4}}{2} = -2 \pm i$$

Complementary function: $x_c = e^{-2t}(A\cos t + B\sin t)$

Particular integral: try $x = p\sin t + q\cos t$

$$\frac{dx}{dt} = p\cos t - q\sin t$$

$$\frac{d^2x}{dt^2} = -p\sin t - q\cos t$$

$\therefore (-p\sin t - q\cos t) + 4(p\cos t - q\sin t)$
$\qquad\qquad + 5(p\sin t + q\cos t) = 8\sin t$

$(-p - 4q + 5p)\sin t + (-q + 4p + 5q)\cos t = 8\sin t$

Equating coefficients of $\cos t$:  $4p + 4q = 0$  $p + q = 0$

Equating coefficients of $\sin t$:  $-4q + 4p = 8$  $p + q = 2$

$\therefore p = 1$  $q = -1$

So the particular integral is $x = \sin t - \cos t$ and the complete
solution is $x = e^{-2t}(A\cos t + B\sin t) + \sin t - \cos t$

> Use the initial conditions given in the question to obtain values for A and B.

$t = 0, x = 5$  $\therefore 5 = A - 1$  $A = 6$

$x = e^{-2t}(A\cos t + B\sin t) + \sin t - \cos t$

$$\frac{dx}{dt} = -2e^{-2t}(A\cos t + B\sin t) + e^{-2t}(-A\sin t + B\cos t) + \cos t + \sin t$$

$t = 0, \dfrac{dx}{dt} = 2$  $\therefore 2 = -2A + B + 1$

$\qquad\qquad B = 2 + 12 - 1 = 13$

So $x = e^{-2t}(6\cos t + 13\sin t) + \sin t - \cos t$

c As $t \to \infty$, $e^{-2t} \to 0$

$\therefore x = \sin t - \cos t$

Write $x = r\sin(t - \alpha)$

$x = r(\sin t \cos \alpha - \cos t \sin \alpha)$

$r\cos \alpha = 1 \quad r\sin \alpha = 1$

$r^2(\cos^2 \alpha + \sin^2 \alpha) = 2 \quad r = \sqrt{2}$

$\tan \alpha = 1 \quad \alpha = \dfrac{\pi}{4}$

$\therefore x = \sqrt{2} \sin\left(t - \dfrac{\pi}{4}\right)$

The motion is S.H.M with amplitude $\sqrt{2}$ and period $2\pi$.

## Example 6

One end of a light elastic string of natural length $l$ and modulus of elasticity $3mk^2l$ is attached to a fixed point $A$ of a horizontal table. A particle $P$ of mass $m$ is attached to the free end of the string and at time $t = 0$ lies at rest on the table with $AP = l$. A force of magnitude $mk^2le^{-2kt}$ acting in the direction $AP$ is applied to $P$. There is a resistance of magnitude $4mkv$, where $v$ is the speed of $P$, acting on $P$. At time $t$, the extension of the string is $x$.

**a** Show that $\dfrac{d^2x}{dt^2} + 4k\dfrac{dx}{dt} + 3k^2x = k^2le^{-2kt}$

**b** Find an expression for $x$ in terms of $k$, $l$ and $t$.

**c** Find the maximum extension of the string.

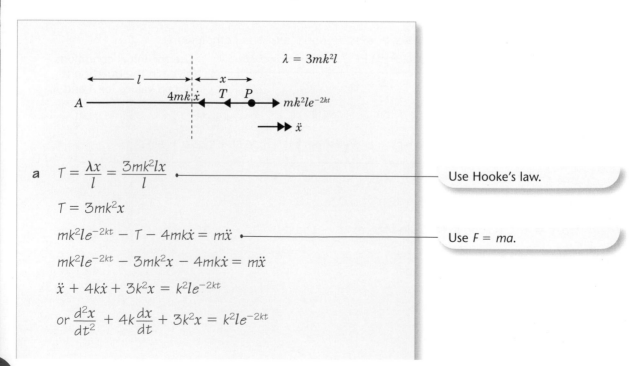

**a** $T = \dfrac{\lambda x}{l} = \dfrac{3mk^2lx}{l}$             Use Hooke's law.

$T = 3mk^2x$

$mk^2le^{-2kt} - T - 4mk\dot{x} = m\ddot{x}$             Use $F = ma$.

$mk^2le^{-2kt} - 3mk^2x - 4mk\dot{x} = m\ddot{x}$

$\ddot{x} + 4k\dot{x} + 3k^2x = k^2le^{-2kt}$

or $\dfrac{d^2x}{dt^2} + 4k\dfrac{dx}{dt} + 3k^2x = k^2le^{-2kt}$

**b** Auxiliary equation: $m^2 + 4km + 3k^2 = 0$

$$(m + 3k)(m + k) = 0$$

$$m = -k \text{ or } -3k$$

Solve the differential equation using the methods of book FP2, Chapter 5.

Complementary function: $x = Ae^{-kt} + Be^{-3kt}$

Particular integral: try $x = ae^{-2kt}$

$\dot{x} = -2kae^{-2kt}, \quad \ddot{x} = 4k^2ae^{-2kt}$

$\therefore 4k^2ae^{-2kt} + 4k(-2kae^{-2kt}) + 3k^2(ae^{-2kt}) = k^2le^{-2kt}$

$$4k^2a - 8k^2a + 3k^2a = k^2l$$

$e^{-2kt} \neq 0$

$$ak^2 = -k^2l$$

$$a = -l$$

Complete solution: $x = Ae^{-kt} + Be^{-3kt} - le^{-2kt}$

$t = 0 \quad x = 0 \Rightarrow 0 = A + B - l$ ①

$\dot{x} = -kAe^{-kt} - 3kBe^{-3kt} + 2kle^{-2kt}$

$t = 0 \quad \dot{x} = 0 \Rightarrow 0 = -kA - 3kB + 2kl$

$$0 = A + 3B - 2l$$ ②

Use the initial conditions given in the question to obtain values for $A$ and $B$.

$\therefore 2B - l = 0 \quad B = \frac{1}{2}l$

$A = -B + l = \frac{1}{2}l$

$\therefore x = \frac{1}{2}le^{-kt} + \frac{1}{2}le^{-3kt} - le^{-2kt}$

**c** When extension is maximum $\dot{x} = 0$.

$\dot{x} = 2kle^{-2kt} - \frac{kl}{2}e^{-kt} - \frac{3kl}{2}e^{-3kt}$

$0 = \frac{kl}{2}e^{-3kt}(4e^{kt} - e^{2kt} - 3)$

$4e^{kt} - e^{2kt} - 3 = 0$

$e^{2kt} - 4e^{kt} + 3 = 0$

$(e^{kt} - 3)(e^{kt} - 1) = 0$

$e^{kt} = 3 \quad t = \frac{1}{k}\ln 3$

$e^{kt} = 1 \quad t = 0 \quad$ (and $x = 0$, so not maximum)

$\therefore x$ is maximum when $t = \frac{1}{k}\ln 3$

$x_{max} = l(\frac{1}{2} \times \frac{1}{3} + \frac{1}{2} \times \frac{1}{27} - \frac{1}{9})$

$e^{kt} = 3$

$= \frac{2}{27}l$

The maximum extension is $\frac{2}{27}l$.

## Example 7

A particle $P$ of mass $m$ is attached to one end of a light elastic string $AB$ of natural length $l$ and modulus of elasticity $mk^2l$. Initially the particle and the string lie at rest on a smooth horizontal plane with $AB = l$. At time $t = 0$ the end $B$ of the spring is set in motion and moves at a constant speed $U$ in the direction $AB$. The air resistance acting on $P$ has magnitude $2mkv$, where $v$ is the speed of $P$. At time $t$ the extension of the spring is $x$ and the displacement of $P$ from its initial position is $y$. Show that, while the string is taut

**a** $x + y = Ut$

**b** $\dfrac{d^2x}{dt^2} + 2k\dfrac{dx}{dt} + k^2x = 2kU.$

**c** Find an expression for $x$ in terms of $U$, $k$ and $t$.

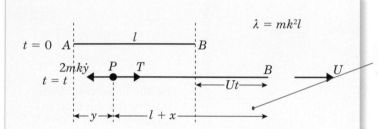

$\lambda = mk^2l$

Draw diagrams showing clearly the situations when $t = 0$ and $t = t$.

**a**  At time $t$ end $B$ has moved a distance $Ut$

$\therefore l + Ut = y + l + x$

$x + y = Ut$ ① 

Use your diagrams to form this equation.

**b**  $T = \dfrac{\lambda x}{l}$

Use Hooke's law.

$T = \dfrac{mk^2lx}{l} = mk^2x$

For $P$: $T - 2mk\dot{y} = m\ddot{y}$

Use $F = ma$.

From ① $\dot{x} + \dot{y} = U \qquad \Rightarrow \dot{y} = U - \dot{x}$

$\qquad\qquad \ddot{x} + \ddot{y} = 0 \qquad \Rightarrow \ddot{y} = -\ddot{x}$

A differential equation with $\dot{x}$ and $\ddot{x}$ is needed.

$\therefore mk^2x - 2mk(U - \dot{x}) = -m\ddot{x}$

$\qquad \ddot{x} + 2k\dot{x} + k^2x = 2kU$

or $\dfrac{d^2x}{dt^2} + 2k\dfrac{dx}{dt} + k^2x = 2kU$

**c**  Auxiliary equation: $m^2 + 2km + k^2 = 0$

Solve the differential equation using the methods of book FP2, Chapter 5.

$\qquad\qquad\qquad (m + k)^2 = 0$

$\qquad\qquad\qquad\qquad m = -k$

Complementary function: $x = (A + Bt)e^{-kt}$

Particular integral: try $x = a$

$$\dot{x} = \ddot{x} = 0$$

$\therefore k^2 a = 2kU \quad a = \dfrac{2U}{k}$

Complete solution: $x = (A + Bt)e^{-kt} + \dfrac{2U}{k}$

$t = 0 \quad x = 0 \quad \Rightarrow 0 = A + \dfrac{2U}{k}$

$\dot{x} = -k(A + Bt)e^{-kt} + Be^{-kt}$

$t = 0 \quad \dot{y} = 0 \quad \Rightarrow \dot{x} = U$

$U = -kA + B$

$B = U + kA = -U$

$\therefore x = \left(-\dfrac{2U}{k} - Ut\right)e^{-kt} + \dfrac{2U}{k}$

Use the initial conditions given in the question to obtain values for $A$ and $B$.

## Exercise 4B

1 A particle $P$ is attached to end $A$ of a light elastic spring $AB$. The end $B$ of the spring is oscillating. At time $t$ the displacement of $P$ from a fixed point is $x$. When $t = 0$, $x = 0$ and $\dfrac{dx}{dt} = \dfrac{k}{5}$ where $k$ is a constant. Given that $x$ satisfies the differential equation $\dfrac{d^2x}{dt^2} + 9x = k\cos t$, find $x$ as a function of $t$.

2 A particle $P$ of mass $m$ lies at rest on a horizontal table attached to end $A$ of a light elastic spring $AB$ of natural length $l$ and modulus of elasticity $9m\omega^2 l$. At time $t = 0$, $AB = l$. The end $B$ of the spring is now moved along the table in the direction $AB$ with constant speed $V$. The resistance to motion of $P$ has magnitude $6m\omega v$, where $v$ is the speed of $P$ and $\omega$ is a constant. At time $t$ the extension of the spring is $x$ and the displacement of $P$ from its initial position is $y$.

Show that

a $x + y = Vt$,

b $\dfrac{d^2x}{dt^2} + 6\omega \dfrac{dx}{dt} + 9\omega^2 x = 6\omega V$.

c Find an expression for $x$ in terms of $t$, $\omega$ and $V$.

3 A particle $P$ of mass $m$ is attached to end $A$ of a light elastic spring $AB$ of natural length $l$ and modulus of elasticity $6mk^2 l$. Initially the spring and the particle lie at rest on a horizontal surface with $AB = l$. The end $B$ of the spring is then moved in a straight line in the direction $AB$ with constant speed $U$. As $P$ moves on the surface it is subject to a resistance of magnitude $5mkv$ where $v$ is the speed of $P$. At time $t$, $t > 0$, the extension of the spring is $x$.

a Show that $\dfrac{d^2x}{dt^2} + 5k\dfrac{dx}{dt} + 6k^2 x = 5kU$.

b Find an expression for $x$ in terms of $t$.

**4** A particle $P$ of mass $m$ is attached to one end of a light elastic string of natural length $a$ and modulus of elasticity $16mn^2a$. The other end of the string is attached to a fixed point $A$ on the horizontal table on which $P$ lies. At time $t = 0$, $P$ is at rest on the table with $AP = a$. A force of magnitude $mn^2ae^{-nt}$, $t \geqslant 0$, acting in the direction $AP$ is applied to $P$. The motion of $P$ is opposed by a resistance of magnitude $10mnv$, where $v$ is the speed of $P$. At time $t$, $t > 0$, the extension of the string is $x$.

  **a** Show that $\dfrac{d^2x}{dt^2} + 10n\dfrac{dx}{dt} + 16n^2x = n^2ae^{-nt}$.

  **b** Find an expression for $x$ in terms of $t$.

**5** A particle $P$ of mass $0.5\,$kg is attached to end $A$ of a light elastic string $AB$ of natural length $0.8\,$m and modulus of elasticity $5\,$N. The particle and string lie on a smooth horizontal plane with $AB = 0.8\,$m. At time $t = 0$ a variable force $F\,$N is applied to the end $B$ of the string which then moves with a constant speed $5\,$m s$^{-1}$ in the direction $AB$. The particle moves along the plane and is subject to air resistance of magnitude $0.5v$ newtons, where $v\,$m s$^{-1}$ is the speed of $P$. At time $t$ seconds the displacement of $P$ from its initial position is $y$ metres and the extension of the string is $x$ metres.

Show that, while the string is taut,

  **a** $x + y = 5t$,

  **b** $\dfrac{d^2x}{dt^2} + \dfrac{dx}{dt} + 12.5x = 5$.

Find

  **c** an expression for $x$ in terms of $t$,

  **d** the exact distance travelled by $P$ in the first $\pi$ seconds,

  **e** the exact value of $F$ when $t = \pi$.

## Mixed exercise 4C

**1** A particle $P$ of mass $0.5\,$kg is free to move horizontally inside a smooth cylindrical tube. The particle is attached to one end of a light elastic spring of natural length $0.2\,$m and modulus of elasticity $5\,$N. At time $t = 0$ the system is at rest with the spring at its natural length. The other end of the spring is then forced to oscillate with simple harmonic motion so that at time $t$ seconds, $t > 0$, its displacement from its initial position is $\frac{1}{5}\sin 2t$ metres and the displacement of $P$ from its initial position is $x$ metres.

  **a** Show that $\dfrac{d^2x}{dt^2} + 50x = 10\sin 2t$.

  **b** Find an expression for $x$ in terms of $t$.

**2** A particle $P$ of mass $m$ is moving in a straight line. At time $t$ the displacement of $P$ from a fixed point $O$ of the line is $x$. Given that $x$ satisfies the differential equation

$$\dfrac{d^2x}{dt^2} + 2k\dfrac{dx}{dt} + n^2x = 0$$

where $k$ and $n$ are positive constants with $k < n$,

  **a** find an expression for $x$ in terms of $k$, $n$ and $t$.

  **b** Write down the period of the motion.

**3** A particle $P$ of mass $m$ is attached to one end of light elastic spring of natural length $l$ and modulus of elasticity $2mk^2l$. The other end of the spring is attached to a fixed point $A$ and $P$ is hanging in equilibrium with $AP$ vertical.

**a** Find the length of the spring.

The particle is now projected vertically downwards from its equilibrium position with speed $U$. A resistance of magnitude $2mkv$, where $v$ is the speed of $P$, acts on $P$. At time $t$, $t > 0$, the displacement of $P$ from its equilibrium position is $x$.

**b** Show that $\dfrac{d^2x}{dt^2} + 2k\dfrac{dx}{dt} + 2k^2x = 0$.

**c** Show that $P$ is instantaneously at rest when $kt = (n + \frac{1}{4})\pi$, where $n \in \mathbb{N}$

**d** Sketch the graph of $x$ against $t$.

**4** A particle $P$ of mass $m$ is attached to one end of a light elastic spring of natural length $l$ and modulus of elasticity $mn^2l$. The other end of the spring is attached to the roof of a stationary lift. The particle is hanging in equilibrium with the spring vertical. At time $t = 0$ the lift starts to move vertically upwards with constant speed $U$. At time $t$, $t > 0$, the displacement of $P$ from its initial position is $x$.

By considering the extension in the spring,

**a** show that $\dfrac{d^2x}{dt^2} + n^2x = n^2Ut$,

**b** find an expression for $x$ in terms of $t$ and $n$.

At time $t = T$, the particle is instantaneously at rest. Find

**c** the smallest value of $T$,

**d** the displacement of $P$ from its initial position at this time.

**5** A particle $P$ of mass $0.6\,\text{kg}$ is attached to one end of light elastic spring of natural length $0.5\,\text{m}$ and modulus of elasticity $1.8\,\text{N}$. The other end of the spring is attached to a fixed point $O$ of the horizontal table on which $P$ lies. At time $t = 0$, $P$ is at the point $A$, where $OA = 0.5\,\text{m}$. The particle is then projected in the direction $OA$ with speed $6\,\text{m\,s}^{-1}$. The particle is subject to a resistance of magnitude $1.2v\,\text{N}$, where $v\,\text{m\,s}^{-1}$ is the speed of $P$. At time $t$ seconds the extension in the spring is $x$ metres.

**a** Show that $\dfrac{d^2x}{dt^2} + 2\dfrac{dx}{dt} + 6x = 0$.

**b** Find $x$ in terms of $t$.

**c** Find the value of $t$ the first time $P$ comes to instantaneous rest.

**6** A particle $P$ of mass $m$ is attached to one end of each of two identical elastic strings of natural length $l$ and modulus of elasticity $2mg$. The free ends of the strings are fixed at points $A$ and $B$ on a smooth horizontal plane where $AB = 4l$. At time $t = 0$, $P$ is at rest at its equilibrium position. The particle is then projected along the line $AB$ with speed $U$ and moves in a straight line. At time $t$ the displacement of $P$ from its equilibrium position is $x$.

A resistance of magnitude $mkv$, where $v$ is the speed of $P$ and $k = \sqrt{\dfrac{g}{l}}$, acts on $P$. Both strings remain taut throughout the motion.

**a** Show that $\dfrac{d^2x}{dt^2} + k\dfrac{dx}{dt} + 4k^2x = 0$.

**b** Find an expression for $x$ in terms of $U$, $k$, and $t$.

## Summary of key points

1  When the simple harmonic motion model for a moving particle is refined to take account of a resistance which is proportional to the speed of the particle, the equation of motion becomes

$$\ddot{x} + k\dot{x} + \omega^2 x = 0$$

2  When an additional force which is a function of time is applied to the particle the equation of motion becomes

$$\ddot{x} + k\dot{x} + \omega^2 x = f(t)$$

where the applied force is $mf(t)$ and $m$ is the mass of the particle.

3  The method for solution follows the same pattern in all cases.

- Obtain the equation of motion for the particle. This will be a second order differential equation as shown above.

- Solve the equation, using the methods of book FP2, Chapter 5.

- Use the information in the question to obtain the arbitrary constants included in the general solution.

After completing this chapter you should be able to:

- find positions of equilibrium in a mechanical system by finding stationary values of the potential energy of the system
- establish whether the equilibrium is stable or unstable.

# Stability

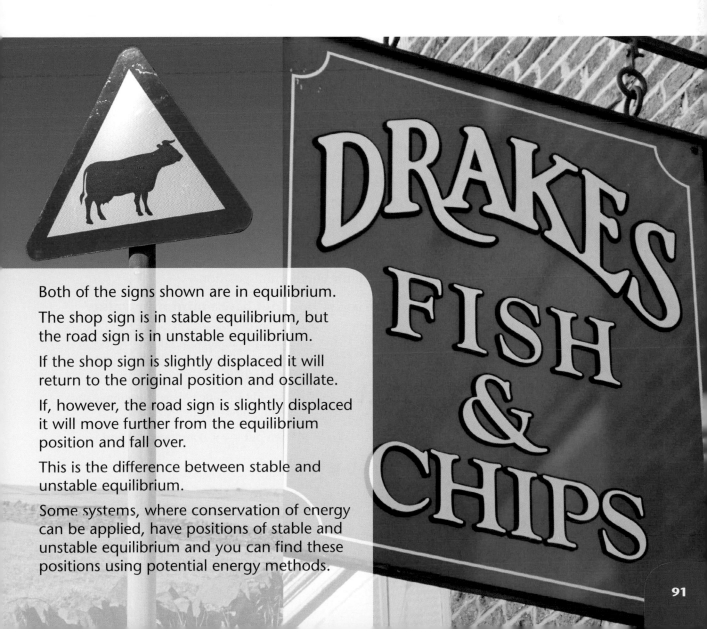

Both of the signs shown are in equilibrium.

The shop sign is in stable equilibrium, but the road sign is in unstable equilibrium.

If the shop sign is slightly displaced it will return to the original position and oscillate.

If, however, the road sign is slightly displaced it will move further from the equilibrium position and fall over.

This is the difference between stable and unstable equilibrium.

Some systems, where conservation of energy can be applied, have positions of stable and unstable equilibrium and you can find these positions using potential energy methods.

## 5.1 You can use potential energy to find positions of equilibrium and to test for stability.

■ If the total potential energy of a conservative system is *V* and the total kinetic knergy of the system is *W* then

$$V + W = k, \text{ a constant,}$$

(i.e. total energy is conserved).

Differentiating with respect to *t* gives:

$$\frac{dV}{dt} + \frac{dW}{dt} = 0 \qquad *$$

But kinetic energy $W = \frac{1}{2}mv^2$, so $\frac{dW}{dt} = mv\frac{dv}{dt}$.

At the positions of equilibrium there is no acceleration so $\frac{dv}{dt} = 0$, and $\frac{dW}{dt} = 0$.

Substitute this into equation *

$$\therefore \frac{dV}{dt} = 0 \text{ at positions of equilibrium.}$$

So if *V* can be expressed in terms of a single variable, frequently $\theta$, and if $\frac{dV}{d\theta} = 0$, then

as $\frac{dV}{dt} = \frac{dV}{d\theta} \cdot \frac{d\theta}{dt}$ this implies $\frac{dV}{dt} = 0$ and so this gives positions of equilibrium.

■ If $\frac{dV}{d\theta} = 0$ then the system is in equilibrium.

A position of **stable** equilibrium is one where the system will move back towards the equilibrium state if it is slightly displaced from the equilibrium position.

When equilibrium is **stable** the potential energy has a **minimum value.**

A position of **unstable** equilibrium is one where the system will move away from the equilibrium state if it is slightly displaced from the equilibrium position.

When equilibrium is **unstable** the potential energy **does not** have a **minimum value.**

■ **Energy condition for equilibrium**

In a mechanical system, which is free to move, and to which the conservation of energy can be applied, positions of equilibrium occur when the potential energy has stationary values.

Express the *total* potential energy *V* of the system is terms of a single variable, frequently $\theta$.

*V* should include both gravitational and elastic potential energy.

Then put $\frac{dV}{d\theta} = 0$ and solve to find $\theta$.

■ **Energy condition for stability**

If the potential energy has a minimum value then the equilibrium is stable.

If the potential energy has a maximum value then the equilibrium is unstable.

If $\frac{d^2V}{d\theta^2} > 0$ then the equilibrium is stable.

If $\frac{d^2V}{d\theta^2} < 0$ then the equilibrium is unstable.

## Example 1

**a** A particle $P$ of mass $m$ rests at the lowest point inside a fixed hollow sphere centre $O$ and radius $r$. By considering the potential energy of the system show that the particle rests in stable equilibrium.

**b** The same particle $P$ now rests on the top of the same fixed sphere. By considering the potential energy of the new system show that the particle now rests in unstable equilibrium

**a**

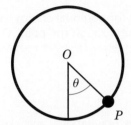

Draw a diagram showing the particle displaced so that $OP$ makes an angle $\theta$ with the vertical.

The potential energy equation is

$V = -mgr\cos\theta + \text{constant}$

So $\dfrac{dV}{d\theta} = mgr\sin\theta$

So for equilibrium $mgr\sin\theta = 0$

i.e. $\theta = 0$

$\dfrac{d^2V}{d\theta^2} = mgr\cos\theta > 0$ when $\theta = 0$

So the equilibrium is stable.

The potential energy is purely gravitational and is clearly at a minimum value when the particle is at the base of the sphere and so the particle rests in stable equilibrium.

Let $V$ be the potential energy in the general position shown.

If you take the base line for potential energy as the horizontal diameter of the sphere then the constant in the formula is zero. If, however, you take the base line as the horizontal plane touching the base of the sphere, then the constant is $mgr$.

Differentiate $V$ and put $\dfrac{dV}{d\theta} = 0$ then solve to obtain $\theta = 0$, which confirms that the lowest point is a point of equilibrium.

Differentiate again. As $\dfrac{d^2V}{d\theta^2} > 0$ when $\theta = 0$ then the equilibrium is stable at the lowest point.

**b**

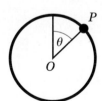

Show the particle displaced so that $OP$ makes an angle $\theta$ with the vertical.

The potential energy equation is

$V = mgr\cos\theta + \text{constant}$

So $\dfrac{dV}{d\theta} = -mgr\sin\theta$

So for equilibrium $mgr\sin\theta = 0$

$\dfrac{d^2V}{d\theta^2} = -mgr\cos\theta = -mgr < 0$

when $\theta = 0$

So the equilibrium is unstable.

The potential energy has a maximum value when the particle is at the top of the sphere and so the particle is not in stable equilibrium.

Let $V$ be the potential energy in the general position shown. The constant depends on the horizontal plane chosen as zero potential energy. If the horizontal plane through $O$ is chosen the constant is zero.

Differentiate $V$ and put $\dfrac{dV}{d\theta} = 0$ then solve to obtain $\theta = 0$.

As $\dfrac{d^2V}{d\theta^2} < 0$ then the equilibrium is unstable.

## Example 2

A particle of mass $m$, is attached at one end of a light elastic string of natural length $l$ and modulus of elasticity $\lambda$. The other end of the string is attached to a fixed point $O$.

**a** Express the potential energy $V$ of the system in terms of $x$, the extension of the string.

**b** Find the value of $x$ when the system is in equilibrium and determine whether this is stable or unstable equilibrium.

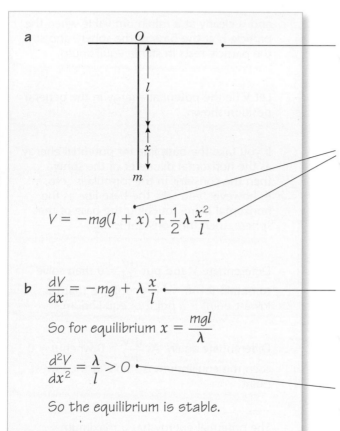

**a**

$$V = -mg(l + x) + \frac{1}{2}\lambda \frac{x^2}{l}$$

Take the base level for potential energy as the horizontal plane through the point $O$.

Take the sum of the gravitational and elastic potential energies.

**b** $\dfrac{dV}{dx} = -mg + \lambda \dfrac{x}{l}$

So for equilibrium $x = \dfrac{mgl}{\lambda}$

$\dfrac{d^2V}{dx^2} = \dfrac{\lambda}{l} > 0$

So the equilibrium is stable.

Differentiate $V$ and put $\dfrac{dV}{dx} = 0$ to obtain $x = \dfrac{mgl}{\lambda}$. (You could alternatively obtain this answer by resolving forces and equating the weight $mg$ with the tension $\dfrac{\lambda x}{l}$. This is the approach used in earlier modules.)

Differentiate again. As $\dfrac{d^2V}{dx^2} > 0$ then the equilibrium is stable.

## Example 3

A bead $B$ of mass $m$ is threaded on a smooth circular wire of radius $a$, which is fixed in a vertical plane. The centre of the circle is $O$. An elastic string $AB$ of natural length $a$ and modulus of elasticity $3mg$ is attached to the bead at one end and to the highest point of the circular wire $A$ at the other end. $AB$ makes an angle $\theta$ with the downward vertical through $A$.

**a** Express the potential energy $V$ of the system in terms of $\theta$.

**b** Find the values of $\theta$, when the system is in equilibrium.

**c** Determine whether they correspond to stable or unstable equilibrium.

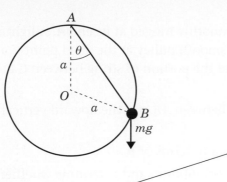

**a**    $V = \frac{1}{2}3mg\dfrac{x^2}{a} - mg(a + x)\cos\theta$, where $x$ is

the extension of the string.

But from triangle $OAB$,

$$a + x = 2a\cos\theta$$

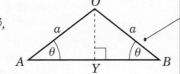

So $V = \dfrac{1}{2}3mg\dfrac{(2a\cos\theta - a)^2}{a} - mg.2a\cos^2\theta$

$V = 4mga\cos^2\theta - 6mga\cos\theta + \frac{3}{2}mga$

**b**    $\dfrac{dV}{d\theta} = -8mga\cos\theta\sin\theta + 6mga\sin\theta$

Put $\dfrac{dV}{d\theta} = 0$, then $mga\sin\theta(6 - 8\cos\theta) = 0$

So $\sin\theta = 0$ or $\cos\theta = \dfrac{3}{4}$

$\theta = 0$ or $\pm 0.723$ radians

**c**    As $\dfrac{dV}{d\theta} = -4mga\sin 2\theta + 6mga\sin\theta$

Then $\dfrac{d^2V}{d\theta^2} = -8mga\cos 2\theta + 6mga\cos\theta$

When $\theta = 0$, $\dfrac{d^2V}{d\theta^2} = -2mga$, which is $< 0$ and

indicates unstable equilibrium at $\theta = 0$.

When $\theta = \pm 0.723$, $\dfrac{d^2V}{d\theta^2} = \dfrac{14}{4}mga$,

which is $> 0$ so indicates stable equilibrium

at these points.

## Example 4

A uniform rod of mass $2m$ and length $2l$ is smoothly hinged at its end $A$. A light inextensible string is attached to the end $B$, passes over a smooth pulley at the fixed point $C$ and carries a mass $m$ at $D$. The string has total length $a$ and the portion of string between $C$ and $D$ hangs vertically.

The distance from $A$ to $C$ is $2l$ and the angle between $AB$ and the upward vertical is $\theta$, where $0 < \theta \leqslant \pi$.

**a** Express the potential energy $V$ of the system in terms of $\theta$.

**b** Find the values of $\theta$ when the system is in equilibrium and determine whether each value corresponds to stable or unstable equilibrium.

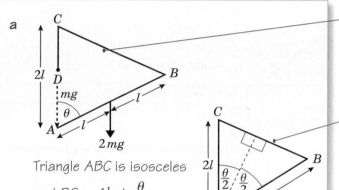

**a**

> Draw a diagram showing the weights acting and their positions below the level of C.

Triangle $ABC$ is isosceles

> Draw the triangle $ABC$ and show its line of symmetry. Use trigonometry on the right angled triangles to find the lengths from $B$ and $C$ to the mid-point of $BC$ and add these to find the length $BC$.

and $BC = 4l \sin \dfrac{\theta}{2}$

So $CD = a - 4l \sin \dfrac{\theta}{2}$

Taking the horizontal through $C$ as the zero level for potential energy:

$$V = -2mg(2l - l\cos\theta) - mg\left(a - 4l\sin\frac{\theta}{2}\right)$$

> Add the two potential energy terms resulting from the mass of $AB$ and from the particle at $D$.

**b** $\dfrac{dV}{d\theta} = -2mgl\sin\theta + 2mgl\cos\dfrac{\theta}{2}$

For equilibrium $\dfrac{dV}{d\theta} = 0$

so $-2mgl\sin\theta + 2mgl\cos\dfrac{\theta}{2} = 0$

$\therefore 2mgl\cos\dfrac{\theta}{2}\left(1 - 2\sin\dfrac{\theta}{2}\right) = 0$

So $\cos\dfrac{\theta}{2} = 0$ or $\sin\dfrac{\theta}{2} = \dfrac{1}{2}$

$\therefore \theta = \pi$ or $\theta = \dfrac{\pi}{3}$

> Differentiate $V$ and put $\dfrac{dV}{d\theta} = 0$ solving the resulting trigonometric equation by using the formula $\sin\theta = 2\sin\dfrac{\theta}{2}\cos\dfrac{\theta}{2}$.

$\dfrac{d^2V}{d\theta^2} = -2mgl\cos\theta - mgl\sin\dfrac{\theta}{2}$

When $\theta = \pi$, $\dfrac{d^2V}{d\theta^2} = +2mgl - mgl > 0$

so equilibrium is stable.

When $\theta = \dfrac{\pi}{3}$, $\dfrac{d^2V}{d\theta^2} = -mgl - \dfrac{1}{2}mgl < 0$

so equilibrium is unstable.

> Differentiate again, checking the sign of $\dfrac{d^2V}{d\theta^2}$ for each of $\theta = \pi$ and $\theta = \dfrac{\pi}{3}$.

> Note: For $\theta = \pi$ the length $CD = a - 4l$ so the condition on $a$ is $a > 4l$ for this configuration to be possible.

> For $\theta = \dfrac{\pi}{3}$ the length $CD = a - 2l$ so the condition on $a$ is $a > 2l$ but $a < 4l$ if $D$ is to be above point $A$, i.e. $2l < a < 4l$ for this configuration to be possible.

## Example 5

A framework consists of two uniform rods $AB$ and $BC$. $AB$ has length $2a$ and mass $m$ and $BC$ has length $4a$ and mass $2m$. The rods are smoothly joined at $B$. The mid-points, $G_1$ and $G_2$, of the rods are joined by a light rod of length $a\sqrt{3}$ so that the angle $ABC$ is 60° or $\frac{\pi}{3}$ radians. The framework is free to rotate in a vertical plane about a fixed smooth horizontal axis through $A$. This axis is perpendicular to the plane of the framework and the angle between $AB$ and the downward vertical through $A$ is denoted by $\theta$.

**a** Express the potential energy $V$ of the system in terms of $\theta$.

**b** Find the values of $\theta$ when the system is in equilibrium and determine whether the equilibrium is stable or unstable in each case.

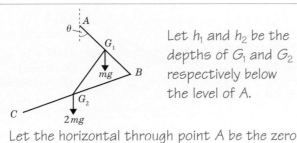

Let $h_1$ and $h_2$ be the depths of $G_1$ and $G_2$ respectively below the level of $A$.

**a** Let the horizontal through point $A$ be the zero level for potential energy, as $A$ is a fixed point

The potential energy of rod $AB$ is $-mga\cos\theta$

The potential energy of rod $BC$ is

$-2mg(2a\cos\theta + 2a\sin\alpha)$, where $\frac{\pi}{3} - \alpha + \theta = \frac{\pi}{2}$

So $\alpha = \theta - \frac{\pi}{6}$

So $V = -mga\cos\theta - 2mg\left(2a\cos\theta + 2a\sin\left(\theta - \frac{\pi}{6}\right)\right)$

i.e. $V = -5mga\cos\theta - 4mga\sin\left(\theta - \frac{\pi}{6}\right)$

**For rod $AB$**
$h_1 = a\cos\theta$.

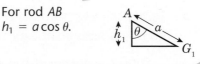

**For rod $BC$**
$h_2 = 2a\cos\theta + 2a\sin\alpha$

and from triangle $ABX$
$\frac{\pi}{3} - \alpha + \theta = \frac{\pi}{2}$

> Work in radians as you are going to differentiate trigonometric functions.

**b** $\frac{dV}{d\theta} = 5mga\sin\theta - 4mga\cos\left(\theta - \frac{\pi}{6}\right)$

For equilibrium $\frac{dV}{d\theta} = 0$, so

$5mga\sin\theta - 4mga\cos\left(\theta - \frac{\pi}{6}\right) = 0$

So $5\sin\theta = 4\cos\theta\cos\frac{\pi}{6} + 4\sin\theta\sin\frac{\pi}{6}$

i.e $5\sin\theta = 2\cos\theta\sqrt{3} + 2\sin\theta$

So $3\sin\theta = 2\cos\theta\sqrt{3}$

and $\tan\theta = \frac{2}{\sqrt{3}}$ so $\theta = 0.857$ or $4.00$ radians

$\frac{d^2V}{d\theta^2} = 5mga\cos\theta + 4mga\sin\left(\theta - \frac{\pi}{6}\right)$

When $\theta = 0.857$, $\frac{d^2V}{d\theta^2} = 4.58$ or $\sqrt{21} > 0$ so stable equilibrium.

> Expand the bracket using $\cos(A - B) = \cos A\cos B + \sin A\sin B$

> Use standard results for $\sin\frac{\pi}{6}$ and $\cos\frac{\pi}{6}$.

> The structure will be below $A$ for the position of stable equilibrium and rotated through $\pi$ above $A$ for the position of unstable equilibrium.

When $\theta = 4.00$, $\frac{d^2V}{d\theta^2} = -4.58$ or $-\sqrt{21} < 0$ so unstable equilibrium.

## Example 6

The figure shows a fixed smooth hemispherical bowl of radius 15 cm, with centre $O$ and with its rim horizontal. A uniform rod $AB$ of length 14 cm rests on the rim of the bowl at $D$ with the end $A$ in contact with the inner surface. The rod has mass $M$ and a particle of mass $M$ is attached to the end $B$. The rod rests in a vertical plane through $O$, and the angle which the rod makes with the horizontal is $\theta$, $0 < \theta < \frac{\pi}{2}$.

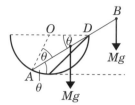

**a** Show that the potential energy of the system $V$ is given by

$$V = -30Mg\sin 2\theta + 21Mg\sin\theta + \text{constant}$$

**b** Find the value of $\cos\theta$ for which the rod is resting in equilibrium and determine whether the equilibrium is stable or unstable.

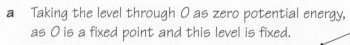

**a** Taking the level through $O$ as zero potential energy, as $O$ is a fixed point and this level is fixed.

$O\hat{D}A = 0$ (alternate angles)
$O\hat{A}D = 0$ (isosceles)
$OZ = 15\sin 2\theta$
$OY = 7\sin\theta$

Potential energy of rod $= -Mg(15\sin 2\theta - 7\sin\theta)$

Potential energy of mass $M$ at point $B$

$= Mg(BD\sin\theta) = Mg(AB - AD)\sin\theta$

$= Mg(14 - 30\cos\theta)\sin\theta$

So total potential energy

$V = -Mg(15\sin 2\theta - 7\sin\theta) + Mg(14 - 30\cos\theta)\sin\theta$

$\quad = -30Mg\sin 2\theta + 21Mg\sin\theta$

> From the isosceles triangle $OAD$
>
> $AD = 2 \times 15\cos\theta = 30\cos\theta$

> Add these two potential energies together to give $V$ and use the double angle formula $\sin 2\theta = 2\sin\theta\cos\theta$ to simplify your expression.

**b** $\dfrac{dV}{d\theta} = -60Mg\cos 2\theta + 21Mg\cos\theta$

Put $\dfrac{dV}{d\theta} = 0$. Then $60Mg\cos 2\theta - 21Mg\cos\theta = 0$

$\therefore 20\cos 2\theta - 7\cos\theta = 0$

So $40\cos^2\theta - 7\cos\theta - 20 = 0$

and $(8\cos\theta + 5)(5\cos\theta - 4) = 0$

So $\cos\theta = \dfrac{4}{5}$ or $\cos\theta = -\dfrac{5}{8}$

But as $0 < \theta < \dfrac{\pi}{2}$, $\cos\theta = \dfrac{4}{5}$ only.

Also $\dfrac{d^2V}{d\theta^2} = 120Mg\sin 2\theta - 21Mg\sin\theta$

When $\cos\theta = \dfrac{4}{5}$, $\sin\theta = \dfrac{3}{5}$ and $\sin 2\theta = \dfrac{24}{25}$

So $\dfrac{d^2V}{d\theta^2} = 120Mg \times \dfrac{24}{25} - 21Mg \times \dfrac{3}{5} = \dfrac{513}{5}Mg > 0$.

So the equilibrium is stable.

> Use the double angle formula
> $\cos 2\theta = 2\cos^2\theta - 1$
> to give a quadratic in $\cos\theta$.

> Solve the quadratic and disregard the negative solution.

> As $\cos\theta = \dfrac{4}{5}$ it is possible to find $\sin\theta$ by using Pythagoras and to find $\sin 2\theta$ by using the double angle formula.

In the above example you could not use the horizontal through $A$ or $B$ as the zero potential energy level. $A$ and $B$ are not fixed points. You could have used the horizontal tangential to the base of the hemisphere as an alternative and this would have given a non-zero constant term in the expression for $V$.

■ **The level of a fixed object must be taken as the zero potential energy level.**

## Exercise 5A

**1** A pendulum is modelled as a uniform rod of mass $m$ and length $2l$ attached to a particle of mass $M$. The pendulum is smoothly hinged at one end to a fixed point $O$, as shown in the figure.

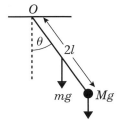

  **a** Express the potential energy of the system in terms of $\theta$, the angle which the pendulum makes with the vertical through $O$.

  **b** Show that there are two positions of equilibrium and determine whether they are stable or unstable.

**2** A small smooth pulley is fixed at a distance $a$ from a fixed smooth vertical wire. A ring of mass $2m$ is free to slide on the wire. It is attached to one end of a string which passes over the pulley and carries a load of mass $4m$ hanging from the other end. The angle between the sloping part of the string and the vertical is $\theta$.

By expressing the potential energy in terms of $\theta$ find how far the ring is below the pulley in the equilibrium position and determine whether the equilibrium is stable or unstable.

**3** The diagram shows a uniform rod $AB$ of length 40 cm and mass $2m$ resting with its end $A$ in contact with a smooth vertical wall. The rod is supported by a smooth horizontal rod which is fixed parallel to the wall and a distance 3 cm from the wall as shown in the figure. A particle of mass $\frac{1}{2}m$ is attached to the rod at $B$.

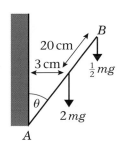

  **a** Show that when $AB$ makes an angle $\theta$ with the vertical the potential energy is given by

  $$V = 0.6mg\cos\theta - 0.075mg\cot\theta + \text{constant}.$$

  **b** Find any positions of equilibrium and establish whether they are stable or unstable.

**4** Two uniform smooth heavy rods, each of mass $M$ and length $2a$, are smoothly jointed together at $B$. They are placed symmetrically in a vertical plane, over a fixed sphere of radius $a$ as shown.

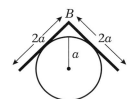

  **a** Show that when the rods make an angle $\theta$ with the horizontal the potential energy $V$ is given by

  $$V = 2Mga(\sec\theta - \sin\theta) + \text{constant}.$$

  > **Hint:** use the horizontal plane through the centre of the sphere as the zero level for the potential energy.

  **b** Show that the rods are in equilibrium if $\cos^3\theta = \sin\theta$ and verify that $\theta = 0.60$ is accurate as a solution to 2 s.f.

**5** Four light rods each of length *l* are freely hinged at their ends to form a rhombus *ABCD* which is suspended from point *A*.

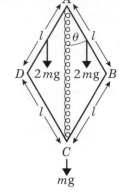

A light spring of natural length *l* and modulus of elasticity 10*mg* connects the points *A* and *C*.

A particle of mass *m* is attached at point *C* and the rods *AB* and *AD* each carry a particle of mass 2*m* at their mid-points. *C* moves freely in a vertical line through *A* and the angle between *AB* and the downward vertical is $\theta$.

  **a** Show that the potential energy of the system *V* is given by

$$V = mgl(20\cos^2\theta - 24\cos\theta) + \text{constant}.$$

  **b** Find the values of $\theta$ which correspond to positions of equilibrium.

  **c** Determine whether these values correspond to stable or to unstable equilibrium.

**6** A light rod *AB* of length 2*a* can turn freely in a vertical plane about a smooth fixed hinge at *A*. A particle of mass *m* is attached at point *B*. One end of a light elastic string, of natural length $\frac{3}{2}a$ and modulus of elasticity $mg\sqrt{3}$ is also attached to the rod at *B*. The other end of the string is attached to a fixed point *O* at the same horizontal level as *A*. Given that *OA* = 2*a* and that the angle between *AB* and the horizontal is 2$\theta$,

  **a** show that, provided the string remains taut, the potential energy of the system is given by

$$V = -2mga(\sin 2\theta + \tfrac{4}{3}\sqrt{3}\cos 2\theta + 2\sqrt{3}\sin\theta) + \text{constant}.$$

  **b** Verify that there is a position of equilibrium in which $\theta = \frac{\pi}{6}$ and determine the stability of this equilibrium.

**7** A small bead *B* of mass *km* can slide on a smooth vertical circular wire with centre *O* and radius *a* which is fixed in a vertical plane. *B* is attached to one end of a light elastic string of natural length $\frac{3}{2}a$ and modulus of elasticity 12*mg*. The other end of the string is attached to a fixed point *A* which is vertically above the point *O* of the circular wire.

The angle between the string *AB* and the downward vertical at *A* is $\theta$.

  **a** Show that the potential energy *V* of the system is given by

$$V = 2mga((8 - k)\cos^2\theta - 12\cos\theta) + \text{constant}.$$

  **b** Find the restrictions on *k* if there is only one point of equilibrium, where $\theta = 0$.

  **c** Subject to these restrictions, determine the stability of this equilibrium.

**8** A uniform rod *AB* of length 2*a* and weight 21*W* is freely pivoted to a fixed support at *A*. A light elastic string of natural length *a* and modulus $\frac{3}{2}W$ has one end attached to *B* and the other to a small ring which is free to slide on a smooth horizontal straight wire passing through a point at a height 9*a* above *A*.

  **a** Show that when the rod makes an angle $\theta$ with the upward vertical at *A* and the string is vertical, the potential energy of the system is

$$V = 3Wa\cos\theta\,(\cos\theta - 1) + \text{constant}.$$

  **b** Find the positions of equilibrium and determine whether they are stable or unstable.

**9** A light rod *AB* can freely turn in a vertical plane about a smooth hinge at *A* and carries a mass *m* hanging from *B*. A light string of length 2*a* fastened to the rod at *B* passes over a smooth peg at a point *C* vertically above *A* and carries a mass *km* at its free end.

If $AC = AB = a$,

**a** find the range of values of *k* for which equilibrium is possible with the rod inclined to the vertical.

**b** Given that equilibrium is possible with the rod horizontal find the value of *k*.

**c** If the rod is slightly disturbed when horizontal and in equilibrium, determine whether it will return to the horizontal position or not.

*E*

**10** Four equal uniform rods, each of length 2*a* and each of mass *M* are rigidly joined together to form a square frame. The frame hangs at rest in a vertical plane on two pegs *P* and *Q* which are at the same level as each other.

If $PQ = b$ and the pegs are each in contact with different rods, show that the potential energy *V* satisfies the equation

$$V = 2Mg(b\sin 2\theta - 2a\sin\theta - 2a\cos\theta).$$

Find the three positions of equilibrium if $b = \sqrt{2}a$ and determine the stability of each of them.

# Summary of key points

**1 Energy condition for equilibrium**

In a mechanical system, which is free to move, and to which the conservation of energy can be applied, positions of equilibrium occur when the potential energy has stationary values.

- Express the *total* potential energy *V* of the system is terms of a single variable, frequently $\theta$.

- The level of a **fixed** object must be taken as the zero potential energy level.

- *V* should include both gravitational and elastic potential energy, where appropriate.

- Then put $\frac{\mathrm{d}V}{\mathrm{d}\theta} = 0$ and solve to find $\theta$.

**2 Energy condition for stability**

If the potential energy has a minimum value then the equilibrium is stable.

If the potential energy has a maximum value (or a point of inflexion) then the equilibrium is unstable.

- If $\frac{\mathrm{d}^2V}{\mathrm{d}\theta^2} > 0$ then the equilibrium is stable.

- If $\frac{\mathrm{d}^2V}{\mathrm{d}\theta^2} < 0$ then the equilibrium is unstable.

# 2 Review Exercise

**1** A particle $P$ moves in a straight line. At time $t$ seconds its displacement from a fixed point $O$ on the line is $x$ metres. The motion of $P$ is modelled by the differential equation

$$\frac{d^2x}{dt^2} + 2\frac{dx}{dt} + 2x = 12\cos 2t - 6\sin 2t.$$

When $t = 0$, $P$ is at rest at $O$.

**a** Find, in terms of $t$, the displacement of $P$ from $O$.

**b** Show that $P$ comes to instantaneous rest when $t = \frac{\pi}{4}$.

**c** Find, in metres to 3 significant figures, the displacement of $P$ from $O$ when $t = \frac{\pi}{4}$.

**d** Find the approximate period of the motion for large values of $t$. **E**

**2**

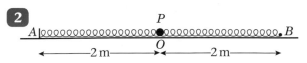

A particle $P$ of mass 2 kg is attached to the mid-point of a light elastic spring of natural length 2 m and modulus of elasticity 4 N. One end $A$ of the elastic spring is attached to a fixed point on a smooth horizontal table. The spring is then stretched until its length is 4 m and its other end $B$ is held at a point on the table where $AB = 4$ m. At time $t = 0$, $P$ is at rest on the table at the point $O$ where $AO = 2$ m, as shown in the diagram. The end $B$ is now moved on the table in such a way that $AOB$ remains a straight line. At time $t$ seconds, $AB = (4 + \frac{1}{2}\sin 4t)$ m and $AP = (2 + x)$ m.

**a** Show that

$$\frac{d^2x}{dt^2} + 4x = \sin 4t.$$

**b** Hence find the time when $P$ first comes to instantaneous rest. **E**

**3**

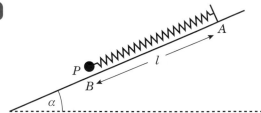

A light elastic spring has natural length $l$ and modulus of elasticity $4mg$. One end of the spring is attached to a point $A$ on a plane that is inclined to the horizontal at an angle $\alpha$, where $\tan \alpha = \frac{3}{4}$. The other end of the spring is attached to a particle $P$ of mass $m$. The plane is rough and the coefficient of friction between $P$ and the plane is $\frac{1}{2}$. The particle $P$ is held at a point $B$ on the plane where $B$ is below $A$ and $AB = l$, with the spring lying along a line of greatest slope of the plane, as shown in the diagram. At time $t = 0$, the particle is projected up the plane towards $A$ with speed $\frac{1}{2}\sqrt{(gl)}$. At time $t$, the compression of the spring is $x$.

**a** Show that
$$\frac{d^2x}{dt^2} + 4\omega^2 x = -g, \text{ where } \omega = \sqrt{\left(\frac{g}{l}\right)}$$

**b** Find $x$ in terms of $l$, $\omega$ and $t$.

**c** Find the distance that $P$ travels up the plane before first coming to rest. **E**

**4** A particle $P$ of mass $m$ is suspended from a fixed point by a light elastic spring. The spring has natural length $a$ and modulus of elasticity $2m\omega^2 a$, where $\omega$ is a positive constant. A time $t = 0$ the particle is projected vertically downwards with speed $U$ from its equilibrium position. The motion of the particle is resisted by a force of magnitude $2m\omega v$, where $v$ is the speed of the particle. At time $t$, the displacement of $P$ downwards from its equilibrium position is $x$.

**a** Show that $\dfrac{d^2x}{dt^2} + 2\omega\dfrac{dx}{dt} + 2\omega^2 x = 0$.

Given that the solution of this differential equation is $x = e^{-\omega t}(A \cos \omega t + B \sin \omega t)$, where $A$ and $B$ are constants,

**b** find $A$ and $B$.

**c** Find an expression for the time at which $P$ first comes to rest. **E**

**5** A light elastic string, of natural length $2a$ and modulus of elasticity $mg$, has a particle $P$ of mass $m$ attached to its mid-point. One end of the string is attached to a fixed point $A$ and the other end is

attached to a fixed point $B$ which is at a distance $4a$ vertically below $A$.

**a** Show that $P$ hangs in equilibrium at the point $E$ where $AE = \frac{5}{2}a$.

The particle $P$ is held at a distance $3a$ vertically below $A$ and is released from rest at time $t = 0$. When the speed of the particle is $v$, there is a resistance to motion of magnitude $2mkv$, where $k = \sqrt{\left(\frac{g}{a}\right)}$. At time $t$ the particle is at a distance $(\frac{5}{2}a + x)$ from $A$.

**b** Show that
$$\frac{d^2x}{dt^2} + 2k\frac{dx}{dt} + 2k^2 x = 0.$$

**c** Hence find $x$ in terms of $t$. **E**

**6** A light spring $PQ$ is fixed at its lower end $Q$ and is constrained to move in a vertical line. At its upper end $P$ the spring is fixed to a small cup, of mass $m$, which contains a sugar lump of mass $m$. The spring has modulus of elasticity $8mg$ and natural length $l$.

Given that the compression of the spring is $x$ at time $t$,

**a** show that, while the sugar lump is in contact with the cup,
$$\frac{d^2x}{dt^2} + \frac{4gx}{l} = g.$$

**b** Given that the system is released from rest when $x = \frac{3l}{4}$ and $t = 0$, show that the lump will lose contact with the cup when $t = \frac{\pi}{3}\sqrt{\frac{l}{g}}$. **E**

**7** A truck is towing a trailer of mass $m$ along a straight horizontal road by means of a tow-rope. The truck and trailer are modelled as particles and the tow-rope is modelled as a light elastic string with modulus of elasticity $4mg$ and natural length $\frac{g}{n^2}$, where $n$ is a positive constant. The effects of friction and air resistance on the trailer are ignored. Initially the trailer is at rest and the tow-rope is slack. The truck then accelerates until the tow-rope is taut and thereafter the truck travels in a straight line with constant speed $u$. At time $t$ after the tow-rope becomes taut, its extension is $x$, and the trailer has moved a distance $y$.

Show that, whilst the rope remains taut,

**a** $y + x = ut$,

**b** $\dfrac{d^2x}{dt^2} + 4n^2x = 0$.

**c** Hence show that the tow-rope goes slack when $t = \dfrac{\pi}{2n}$.

**d** Find the speed of the trailer when $t = \dfrac{\pi}{3n}$.

**e** Find the value of $t$ when the trailer first collides with the truck. **E**

**8** Seats on a coach rest on stabilisers to enable the seats to return to their initial positions smoothly after the coach hits a bump in the road. In a mathematical model of the situation, the following assumptions are made: each stabiliser is a light elastic spring, enclosed in a viscous liquid and fixed in a vertical position; the spring exerts a force of 1.8 N for each cm by which it is extended or compressed; the seat, together with the person sitting on it, constitute a particle $P$ attached to the upper end of the spring which is vertical, the lower end of the spring being fixed; the viscous liquid exerts a resistance to the motion of $P$ of magnitude $240v$ N when the speed of $P$ is $v\,\text{m s}^{-1}$. Given that the mass of $P$ is $m$ kg, and the distance of $P$ from its equilibrium position at time $t$ seconds is $x$ metres measured in a downwards direction,

**a** show that $x$ satisfies the differential equation
$$m\dfrac{d^2x}{dt^2} + 240\dfrac{dx}{dt} + 180x = 0.$$

**b** Show that, when $P$ is disturbed from its equilibrium position, the resulting motion is oscillatory when $m > 80$.

A man is sitting on the seat when the coach hits a bump in the road, giving the seat and initial upward speed of $U\,\text{m s}^{-1}$. The combined mass of the man and the seat is 80 kg.

**c** Find an expression for $x$ in terms of $t$.

**d** Find the greatest displacement of the man from his equilibrium position in the subsequent motion. **E**

**9** A particle $P$ of mass $m$ is attached to the mid-point of a light elastic string, of natural length $2L$ and modulus of elasticity $2mk^2L$, where $k$ is a positive constant. The ends of the string are attached to points $A$ and $B$ on a smooth horizontal surface, where $AB = 3L$. The particle is released from rest at the point $C$, where $AC = 2L$ and $ACB$ is a straight line. During the subsequent motion $P$ experiences air resistance of magnitude $2mkv$, where $v$ is the speed of $P$. At time $t$, $AP = 1.5L + x$.

**a** Show that $\dfrac{d^2x}{dt^2} + 2k\dfrac{dx}{dt} + 4k^2x = 0$.

**b** Find an expression, in terms of $t$, $k$ and $L$, for the distance $AP$ at time $t$. **E**

**10**

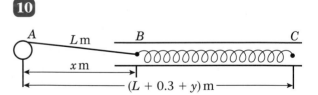

The diagram shows a sketch of a machine component consisting of a long rod $AB$ of length $L$ m. The end $A$ is attached to the circumference of a flywheel centre $O$, radius 0.2 m, which rotates with constant angular speed $10\,\text{rad s}^{-1}$. The other end $B$ is attached to a ring constrained to move in a smooth horizontal tube. The length of the rod is very much greater than the radius of the flywheel and it may be assumed that, at time $t$ seconds, the distance $x$ m, of $B$ from $O$ is given by the equation

$$x = L + 0.2 \cos 10t.$$

Attached to $B$ is a light spring $BC$ of modulus 3.75 N and natural length 0.3 m, at the other end of which is a particle $C$ of mass 0.5 kg which is also constrained to move in the tube. When $t = 0$, the flywheel starts to rotate with $B$ and $C$ at rest and with the spring $BC$ unextended.

**a** Show that, if the distance $OC = (L + 0.3 + y)$ m, then $y$ satisfies the differential equation

$$\frac{\mathrm{d}^2x}{\mathrm{d}t^2} + 25y = 5 \cos 10t.$$

**b** Find an expression for $y$ in terms of $t$. **(E)**

**11** A particle $P$ of mass $m$ kg can move on a smooth horizontal table. It is attached to one end $A$ of an elastic string $AB$, whose natural length is $l$ metres, and whose modulus of elasticity is $10\,mlk^2$ newtons, where $k$ is a positive constant. The string and particle are lying in equilibrium on the table, with $AB = l$ metres. At time $t = 0$, the end $B$ of the string is forced to move horizontally with speed $V \mathrm{m\,s}^{-1}$ in the line of $BA$ and in a direction away from $P$. The end $B$ is forced to maintain this constant speed throughout the subsequent motion. As $P$ moves, it experiences air resistance of magnitude $2mkv$ newtons, where $v \mathrm{m\,s}^{-1}$ is the speed of $P$. After $t$ seconds, the distance of $P$ from its initial position is $x$ metres.

By considering the extension of the string at time $t$,

**a** show that $x$ satisfies the differential equation

$$\frac{\mathrm{d}^2x}{\mathrm{d}t^2} + 2k\frac{\mathrm{d}x}{\mathrm{d}t} + 10k^2x = 10k^2Vt.$$

**b** Find an expression for $x$ in terms of $t$, $k$ and $V$. **(E)**

**12** A particle $P$ of mass $m$ is attached to one end $A$ of a light elastic string $AB$, of natural length $l$ and modulus of elasticity $mln^2$, where $n$ is a constant. The string is lying at rest on a smooth horizontal table, with $AB = l$. At time $t = 0$, the end $B$ is forced to move with constant acceleration $f$ in the direction $AB$ away from $A$. After time $t$, the distance of $P$ from its initial position is $y$, and the extension of the string is $x$.

**a** By finding a relationship between $x$, $y$, $f$ and $t$, show that, while the string remains taut,

$$\frac{\mathrm{d}^2x}{\mathrm{d}t^2} + n^2x = f.$$

**b** Hence express $x$ and $y$ as functions of $t$.

**c** Find the speed of $P$ when the string is at its natural length for the first time in the ensuing motion.

**d** Show that the string never becomes slack. **(E)**

**13** A particle $P$ of mass 0.2 kg is attached to one end of a light elastic string of natural length 0.6 m and modulus of elasticity 0.96 N. The other end of the string is fixed to a point which is 0.6 m above the surface of a liquid. The particle is held on the surface of the liquid, with the string vertical, and then released from rest. The liquid exerts a constant upward force on $P$ of magnitude 1.48 N, and also a resistive force of magnitude $1.2v$ N, when the speed of $P$ is $v \mathrm{m\,s}^{-1}$. At time $t$ seconds, the distance travelled down by $P$ is $x$ metres.

**a** Show that, during the time when $P$ is moving downwards,

$$\frac{\mathrm{d}^2x}{\mathrm{d}t^2} + 6\frac{\mathrm{d}x}{\mathrm{d}t} + 8x = 2.4$$

**b** Find $x$ in terms of $t$.

**c** Show that the particle continues to move down through the liquid throughout the motion. **(E)**

**14** A particle $P$ of mass $m$ is attached to one end of a light elastic string, of natural length $a$ and modulus of elasticity $2mak^2$, where $k$ is a positive constant. The other end of the string is attached to a fixed point $A$. At time $t = 0$, $P$ is released from rest from a point which is a distance $2a$ vertically below $A$. When $P$ is moving with speed $v$, the air resistance has magnitude $2mkv$. At time $t$, the extension of the string is $x$.

**a** Show that, while the string is taut,

$$\frac{d^2x}{dt^2} + 2k\frac{dx}{dt} + 2k^2x = g.$$

You are given that the general solution of this differential equation is

$$x = e^{-kt}(C\sin kt + D\cos kt) + \frac{g}{2k^2}$$

where $C$ and $D$ are constants.

**b** Find the value of $C$ and the value of $D$.

Assuming that the string remains taut,

**c** find the value of $t$ when $P$ first comes to rest,

**d** show that $2k^2a < g(1 + e^{\pi})$. **E**

**15**

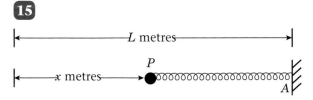

In a simple model of a shock absorber, a particle $P$ of mass $m$ kg is attached to one end of a light elastic horizontal spring. The other end of the spring is fixed at $A$ and the motion of $P$ takes place along a fixed horizontal line through $A$. The spring has natural length $L$ metres and modulus of elasticity $2mL$ newtons. The whole system is immersed in a fluid which exerts a resistance on $P$ of magnitude $3mv$ newtons, where $v\,\text{m s}^{-1}$ is the speed of $P$ at time $t$ seconds. The compression of the spring at time $t$ seconds is $x$ metres, as shown in the diagram.

**a** Show that

$$\frac{d^2x}{dt^2} + 3\frac{dx}{dt} + 2x = 0.$$

Given that when $t = 0$, $x = 2$ and $\frac{dx}{dt} = -4$,

**b** find $x$ in terms of $t$.

**c** Sketch the graph of $x$ against $t$.

**d** State, with a reason, whether the model is realistic. **E**

**16** A light elastic spring, of natural length $a$ and modulus of elasticity $5ma\omega^2$, lies unstretched along a straight line on a smooth horizontal plane. A particle of mass $m$ is attached to one end of the spring. At time $t = 0$, the other end of the spring starts to move with constant speed $U$ along the line of the spring and away from the particle. As the particle moves along the plane it is subject to a resistance of magnitude $2m\omega v$, where $v$ is its speed. At time $t$, the extension of the spring is $x$ and the displacement of the particle from its initial position is $y$. Show that

**a** $x + y = Ut$,

**b** $\dfrac{d^2x}{dt^2} + 2\omega\dfrac{dx}{dt} + 5\omega^2x = 2\omega U.$

**c** Find $x$ in terms of $\omega$, $U$ and $t$. **E**

**17**

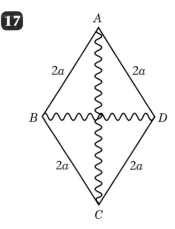

$ABCD$ is a rhombus consisting of four freely jointed uniform rods, each of mass $m$ and length $2a$. The rhombus is freely suspended from $A$ and is prevented from collapsing by two light springs, each of natural length $a$ and modulus of elasticity $2mg$. One spring joins $A$ and $C$ and the other joins $B$ and $D$, as shown in the diagram.

**a** Show that when $AB$ makes an angle $\theta$ with the downward vertical, the potential energy $V$ of the system is given by

$$V = -8mga(\sin\theta + 2\cos\theta) + \text{constant.}$$

**b** Hence find the value of $\theta$, in degrees to one decimal place, for which the system is in equilibrium.

**c** Determine whether this position of equilibrium is stable or unstable.  **E**

**18** A non-uniform rod $BC$ has mass $m$ and length $3l$. The centre of mass of the rod is at distance $l$ from $B$. The rod can turn freely about a fixed smooth horizontal axis through $B$. One end of a light elastic string, of natural length $l$ and modulus of elasticity $\dfrac{mg}{6}$, is attached to $C$. The other end of the string is attached to a point $P$ which is at a height $3l$ vertically above $B$.

**a** Show that, while the string is stretched, the potential energy of the system is

$$mgl(\cos^2\theta - \cos\theta) + \text{constant,}$$

where $\theta$ is the angle between the string and the downward vertical and $-\dfrac{\pi}{2} < \theta < \dfrac{\pi}{2}$.

**b** Find the values of $\theta$ for which the system is in equilibrium with the string stretched.  **E**

**19**

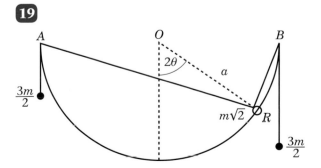

A smooth wire with ends $A$ and $B$ is in the shape of a semi-circle of radius $a$. The mid-point of $AB$ is $O$. The wire is fixed in a vertical plane and hangs below $AB$ which is horizontal. A small ring $R$, of mass $m\sqrt{2}$, is threaded on the wire and is attached to two light inextensible strings. The other end of each string is attached to a particle of mass $\dfrac{3m}{2}$. The particles hang vertically under gravity, as shown in the diagram.

**a** Show that, when the radius $OR$ makes an angle $2\theta$ with the vertical, the potential energy, $V$, of the system is given by

$$V = \sqrt{2}mga(3\cos\theta - \cos 2\theta) + \text{constant,}$$

**b** find the values of $\theta$ for which the system is in equilibrium.

**c** Determine the stability of the position of equilibrium for which $\theta > 0$.  **E**

**20**

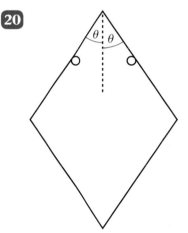

Four equal uniform rods each of length $2a$ and mass $m$ are smoothly jointed to form a rhombus. This is used by a gardener to measure areas of lawn for treatment. When not in use it is stored resting on two smooth pegs, which are at the same level and a distance $\frac{1}{2}a$ apart, with the rhombus in a vertical plane, as shown in the diagram. Given that each of the rods make an angle $\theta$ with the vertical,

**a** show that the potential energy of the system is

$$mga \cot \theta - 8mga \cos \theta + c,$$

where $c$ is a constant.

**b** Hence find the value of $\theta$ when the system is in equilibrium. **E**

**21**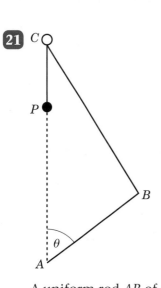

A uniform rod $AB$ of mass $2m$ and length $2a$ is freely hinged about a horizontal axis through $A$. The end $B$ is attached to a light inextensible string of length $b$, $2a < b < 6a$, which passes through a small, smooth ring at $C$. A particle $P$, of mass $m$, is attached to the other end of the string and hangs freely. The point $C$ is vertically above the point $A$ and $AC = 4a$. The angle $CAB$ is denoted by $\theta$, as shown in the diagram.

**a** Show that the total potential energy of the system is given by

$$2mga(\cos \theta + \sqrt{[5 - 4 \cos \theta]} + k),$$

where $k$ is a constant.

**b** Find, in degrees to 1 decimal place, a value of $\theta$, $0 < \theta < 180°$, for which the system is in equilibrium. **E**

**22**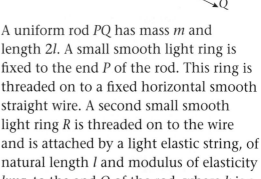

A uniform rod $PQ$ has mass $m$ and length $2l$. A small smooth light ring is fixed to the end $P$ of the rod. This ring is threaded on to a fixed horizontal smooth straight wire. A second small smooth light ring $R$ is threaded on to the wire and is attached by a light elastic string, of natural length $l$ and modulus of elasticity $kmg$, to the end $Q$ of the rod, where $k$ is a constant.

**a** Show that, when the rod $PQ$ makes an angle $\theta$ with the vertical, where $0 < \theta \leqslant \frac{\pi}{3}$, and $Q$ is vertically below $R$, as shown in the diagram, the potential energy of the system is

$$mgl[2k \cos^2 \theta - (2k + 1) \cos \theta]$$
$$+ \text{ constant}.$$

Given that there is a position of equilibrium with $\theta > 0$,

**b** show that $k > \frac{1}{2}$. **E**

**23**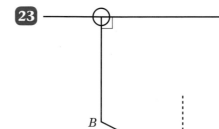

A uniform rod $AB$, of length $2a$ and mass $8m$, is free to rotate in a vertical plane about a fixed smooth horizontal axis through $A$. One end of a light elastic string, of natural length $a$ and modulus of elasticity $\frac{4}{5}mg$, is fixed to $B$. The other end of the string is attached to a small ring which is free to slide on a smooth straight horizontal wire which is fixed in the same vertical plane as $AB$ at a height $7a$ vertically above $A$. The rod $AB$ makes an angle $\theta$ with the upward vertical at $A$, as shown in the diagram.

**a** Show that the potential energy $V$ of the system is given by

$$V = \tfrac{8}{5}mg\,a(\cos^2\theta - \cos\theta) + \text{constant.}$$

**b** Hence find the values of $\theta$, $0 \le \theta \le \pi$, for which the system is in equilibrium.

**c** Determine the nature of these positions of equilibrium. **E**

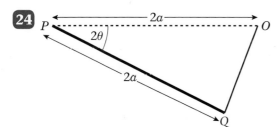

**24** A uniform rod $PQ$, of length $2a$ and mass $m$, is free to rotate in a vertical plane about a fixed smooth horizontal axis through the end $P$. The end $Q$ is attached to one end of a light elastic string, of natural length $a$ and modulus of elasticity $\dfrac{mg}{2\sqrt{3}}$. The other end of the string is attached to a fixed point $O$, where $OP$ is horizontal and $OP = 2a$, as shown in the diagram. $\angle OPQ$ is denoted by $2\theta$.

**a** Show that, when the string is taut, the potential energy of the system is

$$-\frac{mga}{\sqrt{3}}(2\cos 2\theta + \sqrt{3}\sin 2\theta + 2\sin\theta) + \text{constant.}$$

**b** Verify that there is a position of equilibrium at $\theta = \dfrac{\pi}{6}$.

**c** Determine whether this is a position of stable equilibrium. **E**

**25**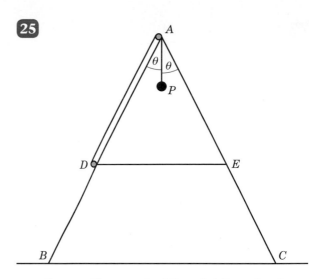

Two uniform rods $AB$ and $AC$, each of mass $2m$ and length $2L$, are freely jointed at $A$. The mid-points of the rods are $D$ and $E$ respectively. A light inextensible string of length $s$ is fixed to $E$ and passes round small, smooth light pulleys at $D$ and $A$. A particle $P$ of mass $m$ is attached to the other end of the string and hangs vertically. The points $A$, $B$ and $C$ lie in the same vertical plane with $B$ and $C$ on a smooth horizontal surface. The angles $PAB$ and $PAC$ are each equal to $\theta$ ($\theta > 0$), as shown in the diagram.

**a** Find the length of $AP$ in terms of $s$, $L$ and $\theta$.

**b** Show that the potential energy $V$ of the system is given by

$$V = 2mgL\,(3\cos\theta + \sin\theta) + \text{constant.}$$

**c** Hence find the value of $\theta$ for which the system is in equilibrium.

**d** Determine whether this position of equilibrium is stable or unstable. **E**

**26**

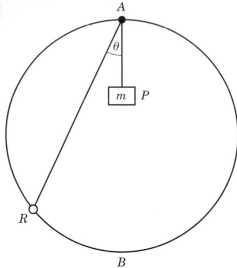

A smooth wire $AB$, in the shape of a circle of radius $r$, is fixed in a vertical plane with $AB$ vertical. A small smooth ring $R$ of mass $m$ is threaded on the wire and is connected by a light inextensible string to a particle $P$ of mass $m$. The length of the string is greater than the diameter of the circle. The string passes over a small smooth pulley which is fixed at the highest point $A$ of the wire and angle $R\hat{A}P = \theta$, as shown in the diagram.

**a** Show that the potential energy of the system is given by

$$2mgr(\cos\theta - \cos^2\theta) + \text{constant.}$$

**b** Hence determine the values of $\theta$, $\theta \geqslant 0$, for which the system is in equilibrium.

**c** Determine the stability of each position of equilibrium.

**27**

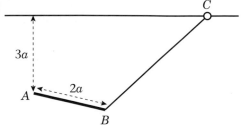

A uniform rod $AB$ has mass $m$ and length $2a$. One end $A$ is freely hinged to a fixed point. One end of a light elastic string, of natural length $a$ and modulus $\frac{1}{2}mg$, is

attached to the other end $B$ of the rod. The other end of the string is attached to a small ring $C$ which can move freely on a smooth horizontal wire fixed at a height of $3a$ above $A$ and in the vertical plane through $A$, as shown in the diagram.

**a** Explain why, when the system is in equilibrium, the elastic string is vertical.

**b** Show that, when $BC$ is vertical and the rod $AB$ makes an angle $\theta$ with the downward vertical, the potential energy, $V$, of the system is given by

$$V = mga(\cos^2\theta + \cos\theta) + \text{constant.}$$

**c** Hence find the values of $\theta$, $0 \leqslant \theta \leqslant \pi$, for which the system is in equilibrium.

**d** Determine whether each position of equilibrium is stable or unstable.

**28**

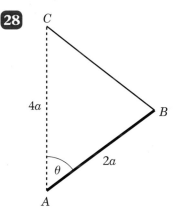

A uniform rod $AB$, of mass $m$ and length $2a$, can rotate freely in a vertical plane about a fixed smooth horizontal axis through $A$. The fixed point $C$ is vertically above $A$ and $AC = 4a$. A light elastic string, of natural length $2a$ and modulus of elasticity $\frac{1}{2}mg$, joins $B$ to $C$. The rod $AB$ makes an angle $\theta$ with the upward vertical at $A$, as shown in the diagram.

**a** Show that the potential energy of the system is

$$-mga[\cos\theta + \sqrt{(5 - 4\cos\theta)}] + \text{constant.}$$

**b** Hence determine the values of $\theta$ for which the system is in equilibrium.

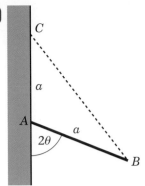

A uniform rod $AB$, of mass $m$ and length $a$, can rotate in a vertical plane about a smooth hinge fixed at $A$ on a vertical wall. A point $C$ on the wall is at a height $a$ vertically above $A$. One end of an elastic string, of natural length $a$ and modulus of elasticity $2mg$, is attached to $C$ and the other end is attached to the end $B$ of the rod, as shown in the diagram.

**a** Show that, when the rod $AB$ makes an angle $2\theta$, $\theta > 0$, with the downward vertical, and the string is taut, the potential energy, $V$, of the system is given by
$$V = -\tfrac{1}{2}mga\cos 2\theta + mga(2\cos\theta - 1)^2 + \text{constant}.$$

**b** Hence determine the value of $\theta$ for which the system is in equilibrium.

**c** Determine whether this position of equilibrium is stable or unstable. **E**

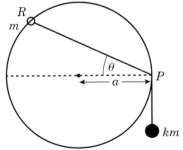

A small ring $R$, of mass $m$, is free to slide on a smooth wire in the shape of a circle with radius $a$. The wire is fixed in a vertical plane. A light inextensible string has one end attached to $R$ and passes over a small smooth pulley at $P$, where

$P$ is one end of the horizontal diameter of the wire. The other end of the string is attached to a mass $km$ ($k < 1$) which hangs freely, as shown in the diagram.

$PR$ makes an angle $\theta$ with the horizontal.

**a** Show that the potential energy of the system, $V$, is given by
$$V = mga(\sin 2\theta + 2k\cos\theta) + \text{constant}.$$

Given that $k = \tfrac{1}{2}$,

**b** find, in radians to 3 decimal places, the values of $\theta$, $-\dfrac{\pi}{2} \leqslant \theta \leqslant \dfrac{\pi}{2}$, for which the system is in equilibrium.

**c** Determine whether each of the positions of equilibrium is stable or unstable. **E**

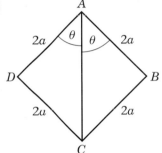

Four identical uniform rods, each of mass $m$ and length $2a$, are freely jointed to form a rhombus $ABCD$. The rhombus is suspended from $A$ and is prevented from collapsing by an elastic string which joins $A$ to $C$, with $\angle BAD = 2\theta$, $0 \leqslant \theta \leqslant \tfrac{1}{3}\pi$, as shown in the diagram. The natural length of the elastic string is $2a$ and its modulus of elasticity is $4mg$.

**a** Show that the potential energy, $V$, of the system is given by
$$V = 4mga\left[(2\cos\theta - 1)^2 - 2\cos\theta\right] + \text{constant}.$$

**b** Hence find the non-zero value of $\theta$ for which the system is in equilibrium.

**c** Determine whether this position of equilibrium is stable or unstable. **E**

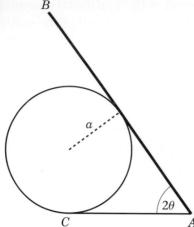

The diagram shows a uniform rod $AB$, of mass $m$ and length $4a$, resting on a smooth fixed sphere of radius $a$. A light elastic string, of natural length $a$ and modulus of elasticity $\frac{3}{4}mg$, has one end attached to the lowest point $C$ of the sphere and the other end attached to $A$. The points $A$, $B$ and $C$ lie in a vertical plane with $\angle BAC = 2\theta$, where $\theta < \frac{\pi}{4}$.

Given that $AC$ is always horizontal,

**a** show that the potential energy of the system is

$$\frac{mga}{8}(16\sin 2\theta + 3\cot^2 \theta - 6\cot \theta)$$
$$+ \text{constant},$$

**b** show that there is a value for $\theta$ for which the system is in equilibrium such that $0.535 < \theta < 0.545$.

**c** Determine whether this position of equilibrium is stable or unstable.   **E**

# Examination style paper

1 A smooth sphere $S$ is moving on a smooth horizontal plane with speed $u$ when it collides with a smooth fixed vertical wall. At the instant of collision the direction of motion of $S$ makes an angle of 30° with the wall. The coefficient of restitution between $S$ and the wall is $\frac{1}{5}$.

Find the speed of $S$ immediately after the collision.

(Total 6 marks)

2 A particle moves along a straight line. At time $t$ the displacement of the particle from a fixed point $O$ is $x$, where

$$\ddot{x} + 2\dot{x} + 3x = 0.$$

a Show that the particle performs damped harmonic motion. (5)

b State the period of the damped harmonic motion. (1)

Given also that when $t = 0$, $x = 0$ and $\dot{x} = 3$,

c find the maximum displacement of the particle from O. (5)

(Total 11 marks)

3 At noon, two yachts $A$ and $B$ are 5 km apart, with $A$ due north of $B$. Yacht $A$ is sailing at a constant speed of $12\,\text{km h}^{-1}$ on a bearing of 120°. Yacht $B$ is moving at a constant speed of $10\,\text{km h}^{-1}$.

a Show that it is not possible for $B$ to intercept $A$. (3)

Yacht $B$ sets a course to pass as close as possible to $A$. Find

b the direction of motion of $B$, giving your answer as a bearing, (4)

c the shortest distance between the two yachts in the subsequent motion if $B$ sets this course. (2)

(Total 9 marks)

4 A body falls vertically from rest and is subject to air resistance of a magnitude which is proportional to the square of its speed.

Given that the terminal speed of the body is $50\,\text{m s}^{-1}$, find the distance fallen in the time it takes for the body to attain a speed of $25\,\text{m s}^{-1}$.

(Total 10 marks)

**5** Two men are on the same road. One is jogging due north at a constant speed $u$ and to him the wind seems to be blowing from the northeast. The other is cycling due south at a constant speed $2u$. To him the wind appears to be blowing from the southeast. Assuming that the velocity, $\mathbf{w}$, of the wind relative to the ground is constant, find

  **a** the magnitude of $\mathbf{w}$, in terms of $u$,            (7)

  **b** the direction of $\mathbf{w}$.            (3)

                                  **(Total 10 marks)**

**6** Two small smooth spheres $A$ and $B$ have equal radii. The mass of $A$ is $2m$ kg and the mass of $B$ is $m$ kg. The spheres are moving on a smooth horizontal plane and they collide. Immediately before the collision the velocity of $A$ is $(4\mathbf{i} + 5\mathbf{j})\,\mathrm{m\,s}^{-1}$ and the velocity of $B$ is $(3\mathbf{i} + 3\mathbf{j})\,\mathrm{m\,s}^{-1}$. Immediately after the collision the velocity of $A$ is $(3\mathbf{i} + 4\mathbf{j})\,\mathrm{m\,s}^{-1}$.

  **a** Find the velocity of $B$ immediately after the collision.     (3)

  **b** Show that the impulse received by $A$ in the collision is $-2m(\mathbf{i} + \mathbf{j})$ Ns.     (3)

  **c** Find the coefficient of restitution between the two spheres.     (8)

                                  **(Total 14 marks)**

**7** A uniform rod $AB$ of length $l$ and mass $m$ has end $A$ freely joined to a fixed point. A light elastic string with modulus of elasticity $2mg$ and natural length $l$ has one end attached to the rod at $B$ and the other attached to point $C$ which lies a distance $l$ vertically above $A$.

  **a** Show that when the rod is at an angle of $2\theta$ to the vertical the potential energy of the system is

$$mgl\left[(2\sin\theta - 1)^2 + \tfrac{1}{2}\cos 2\theta\right] + \text{constant.}$$
                                           (6)

  **b** Hence show that there are two positions in which the rod can rest in equilibrium with the string taut.     (4)

  **c** Determine the stability of the positions of equilibrium.     (5)

                                  **(Total 15 marks)**

# Answers

## Exercise 1A

1. **a** $\sqrt{82}\,\mathrm{m\,s^{-1}}$ **b** $\sqrt{29}\,\mathrm{m\,s^{-1}}$ **c** $\sqrt{185}\,\mathrm{m\,s^{-1}}$
2. $40\,\mathrm{km/h^{-1}}\,\mathrm{N}\,60°\,\mathrm{E}$
3. $(\mathbf{i} + 7\mathbf{j})\,\mathrm{m\,s^{-1}}$
4. $5.05\,\mathrm{km/h^{-1}}\,\mathrm{N}\,41.9°\,\mathrm{E}$
5. $29.7°$
6. $32.68°\,\mathrm{S}$ of $\mathrm{E}$
7. SW
8. 3 h 28 minutes (nearest minute)
9. **a** $\mathrm{S}\,52.8°\,\mathrm{W}$ **b** $632\,\mathrm{km\,h^{-1}}$
10. 12:13
11. **a** $\mathrm{N}\,32.8°\,\mathrm{E}$
    **b** 30 minutes (nearest minute)
13. $\mathrm{N}\,75°\,\mathrm{E}$
14. $18.0\,\mathrm{km\,h^{-1}},\,\mathrm{N}\,60.9°\,\mathrm{E}$
15. $98.5\,\mathrm{km\,h^{-1}},\,\mathrm{S}\,66°\,\mathrm{E}$
16. $34.0\,\mathrm{km\,h^{-1}},\,\mathrm{S}\,74.9°\,\mathrm{E}$

## Exercise 1B

1. **b** 11.06 a.m. **c** $(80\mathbf{i} + 460\mathbf{j})\,\mathrm{km}$
2. **a** $\mathrm{N}\,44.4°\,\mathrm{W}$ **b** 36.7 minutes
3. **a** $\mathrm{N}\,24.7°\,\mathrm{E}$ **b** 2.4 s (1 d.p.)
4. **a** 24.2 km **b** $48.1°$
5. **a** $\mathrm{N}\,15.5°\,\mathrm{W}$ **b** 9.1 minutes (1 d.p.)
6. 10.26 a.m.
7. $\mathrm{N}\,67.3°\,\mathrm{W}$

## Exercise 1C

1. **a** $\sqrt{17}\,\mathrm{km}$ **b** 9.30 a.m.
2. **a** $6\sqrt{5}\,\mathrm{km}$ **b** 9.48 a.m.
3. **a** $\sqrt{17}\,\mathrm{km}$ **b** 4 p.m.
4. **a** $2\sqrt{5}\,\mathrm{km}$ **b** 10 minutes
5. **a** $\sqrt{2}\,\mathrm{km}$ **b** 12 minutes

## Exercise 1D

1. **a** 1 h 4 minutes **b** 6 km
2. **a** $117\,\mathrm{km\,h^{-1}}$ **b** 0.692 km (3 s.f.)
3. **a** 72.1 km **b** 10.24 a.m. (nearest minute)
4. **a** $14.3\,\mathrm{km\,h^{-1}}$ **b** 3.12 km
5. **a** 0.174 km **b** 46 minutes

## Exercise 1E

1. **b** $\mathrm{N}\,6.87°\,\mathrm{W}$ **c** 0.48 km **d** $2.26\frac{1}{2}$ p.m.
2. **a** $108°$ (nearest degree) **b** 1.56 km
   **c** 21.2 s
3. **a** $\mathrm{N}\,22.6°\,\mathrm{E}$ **b** 2.00 km
   **c** 3.58 p.m. (nearest minute) **d** 11.5 km (3 s.f.)
4. **a** 6.2 m **b** 1.96 s
5. **a** $330°$ **b** 8 km **c** 10.24 a.m.
   **d** 20.8 minutes **e** 14.3 minutes

## Exercise 1F

1. **a** $(-2\mathbf{i} + 2\mathbf{j})\,\mathrm{km\,h^{-1}}$ **c** $7\sqrt{2}\,\mathrm{km}$
   **d** $\frac{1}{2}\mathrm{h}$ **f** 9.91 km
2. **a** $\mathrm{N}\,23.6°\,\mathrm{E}$ **b** 13.1 minutes
3. **b** $50\sqrt{2}$ **c** $50\,\mathrm{km\,h^{-1}}$ due N
4. **b** 0.571 km
5. 47 minutes
6. **a** 22.5 minutes **b** 1.89 km
   **c** $12.1\,\mathrm{km\,h^{-1}},\,298°$
7. **a** 0.924 km **b** 0.5 km
8. **a** $14.9\,\mathrm{km\,h^{-1}}$ **b** $\mathrm{N}\,56.25°\,\mathrm{W}$
9. **a** $\mathrm{N}\,34.8°\,\mathrm{E}$ **b** 3 h 6 minutes

## Exercise 2A

1. $v = \dfrac{u\sqrt{17}}{5}$
2. $e = \frac{1}{4}$
3. $v = \dfrac{3\sqrt{17}u}{13}$
4. $e = \dfrac{\sqrt{29}}{8}$
5. $\sqrt{19}\,\mathrm{m\,s^{-1}}$
6. $5.08\,\mathrm{m\,s^{-1}}$
7. **a** $5.59\,\mathrm{m\,s^{-1}}$ **b** $5.625\,\mathrm{N\,s}$
8. $e = 0.36$
9. **a** $v = -2.5\mathbf{i} - 3\mathbf{j}$ **b** 7.5 J
10. **a** $\sqrt{13}\,\mathrm{m\,s^{-1}}$ **b** 16 J
11. **a** $\frac{8}{3}\mathbf{i} + \frac{10}{3}\mathbf{j}$ **b** $\frac{16}{9}$ J
12. $6.09\,\mathrm{m\,s^{-1}}$ at $21.8°$ to the cushion
13. **b** 0.7
14. **a** $\frac{56}{225}$ **b** $\frac{5}{9}$
15. **a** parallel to the original path but in the opposite direction
    **b** $eu$
16. **a** $5m\,\mathrm{N\,s}$ in the direction parallel to the unit vector $\frac{1}{5}(-3\mathbf{i} + 4\mathbf{j})$
    **b** $e = \frac{2}{23}$
17. **a** $2\sqrt{17}\,\mathrm{N\,s}$ in the direction parallel to the unit vector $\frac{1}{\sqrt{17}}(\mathbf{i} - 4\mathbf{j})$
    **b** $e = \frac{7}{10}$ **c** 3 J
18. $\dfrac{\sqrt{3}u}{3}\,\mathrm{m\,s^{-1}}$

## Exercise 2B

1. $1.04\,\mathrm{m\,s^{-1}}$ perpendicular to the line of centres
   $2.94\,\mathrm{m\,s^{-1}}$ parallel to the line of centres

**2** $\frac{4\sqrt{39}}{9}$ m s$^{-1}$ at 46.1° to the line of centres

$\frac{16\sqrt{3}}{9}$ m s$^{-1}$ along the line of centres

**3** $\frac{25}{7}$ m s$^{-1}$ at 81.9° to the line of centres

$\frac{45\sqrt{2}}{28}$ m s$^{-1}$ along the line of centres

**5** $\frac{\sqrt{61}v}{6}$ m s$^{-1}$ at 50.2° to the line of centres

$\frac{1}{6}v$ m s$^{-1}$ along the line of centres

**7** 2.53 J, 3.06 N s

**8 a** $\sqrt{13}$ m s$^{-1}$, $2\sqrt{5}$ m s$^{-1}$    **b** $\frac{10}{43}$

**9** $4\mathbf{i} + \mathbf{j}$ m s$^{-1}$, $2\mathbf{i}$ m s$^{-1}$

**10** 3.23 m s$^{-1}$, 3.25 m s$^{-1}$

**11** 1.92 J

**12 a** $\left(\frac{5}{2}\mathbf{i} + \frac{1}{2}\mathbf{j}\right)$ m s$^{-1}$    **b** $\frac{1}{\sqrt{10}}(\mathbf{i} - 3\mathbf{j})$

**13 a** $\sqrt{5}$ m s$^{-1}$    **b** 9 mJ

**14 a** 0    **b** $\frac{1}{7}$

**16** 1 m s$^{-1}$, $\sqrt{13}$ m s$^{-1}$

## Exercise 2C

**1** $\frac{\sqrt{7}}{5}$

**2 a** 2.33 m s$^{-1}$    **b** 3 N s

**3 a** $\frac{1}{2}\mathbf{i} + \frac{5}{2}\mathbf{j}$    **b** 3.375 J

**4 a** $m\sqrt{41}$ N s parallel to $\frac{1}{\sqrt{41}}(-4\mathbf{i} - 5\mathbf{j})$    **b** $\frac{2}{39}$

**5 a** $-\mathbf{i} + 3\mathbf{j}$    **b** $\frac{\sqrt{2}}{2}(\mathbf{i} - \mathbf{j})$

**8** $\sqrt{5}$ m s$^{-1}$, 1.5 m s$^{-1}$

**10** $\frac{3u}{2}$, $\frac{3\sqrt{29}}{10}u$

## Exercise 3A

**1** $\frac{1}{4}\ln 4$ s = 0.347 s (3 d.p.)

**2** $2 \ln 2$ m = 1.39 m (3 s.f.)

**3 a** $\ln 2.5$ s = 0.916 s (3 d.p.)

   **b** $(12 - 8\ln 2.5)$ m = 4.67 m (3 s.f.)

**4** $\frac{1}{2kg}\ln\left(\frac{\mu + ku^2}{\mu}\right)$

**5 b** $\left(\frac{m}{2k}\right)\ln\left(\frac{4a^2 + 4U^2}{4a^2 + U^2}\right)$

**6 a** $12\left(1 - e^{-\frac{t}{4}}\right)$    **b** 12 m s$^{-1}$

## Exercise 3B

**1** 23.9 m (3 s.f.)

**2** 0.342 (3 s.f.)

**3 b** $\frac{1}{k}\ln\left(1 + \frac{ku}{g}\right)$

**4 a** 18.6 m s$^{-1}$ (3 s.f.)    **b** 2.97 m s$^{-1}$ (3 s.f.)

**5 a** $\frac{1}{gk^2}(ku - \ln(1 + ku))$    **b** $\frac{1}{k}(1 - e^{-kgt})$

**6 a** $\frac{u^2}{2g}$    **b** $\frac{1}{2k}\ln\left(1 + \frac{ku^2}{g}\right)$

   **c** $mg\left(\frac{u^2}{2g} - \frac{1}{2k}\ln\left(1 + \frac{ku^2}{g}\right)\right)$

**7 b** $\frac{g}{8k^2}(16\ln 2 - 3)$

**8** $\frac{1}{2k}\ln\left(1 + \frac{8kU^2}{7g}\right)$

## Exercise 3C

**1** 9.85 s (3 s.f.)

**2** $\frac{1}{2k}\ln\left(\frac{15}{7}\right)$

**3 a** 20 m s$^{-1}$    **b** 33.2 m (3 s.f.)

**4** $\frac{mk^3}{3}\ln\left(\frac{9}{8}\right)$

## Exercise 3D

**1** $\frac{1}{3kU}$

**2** $\left(\frac{g}{k}\right)^{\frac{1}{2}}(1 - e^{-2kD})^{\frac{1}{2}}$

**3** 56.1 m (3 s.f.)

**4** $d = \frac{1}{10b}\ln(1 + bU^2)$

**5 b** $\frac{c}{g}\arctan\left(\frac{V}{c}\right)$

**7 b** 0.24 (2 d.p.)

**8 b** $\frac{1}{k}\ln\left(\frac{21P}{8(3P - mkv^3)}\right)$

## Review Exercise 1

**1 a** $\frac{5}{6}$ m s$^{-1}$    **b** $\frac{4}{3}$ m s$^{-1}$

**2** 096° (nearest degree)

**3 a** 21° (nearest degree)    **b** 3.7 m s$^{-1}$ (2 s.f.)

   **c** 108 s (nearest second)

**4** 151 s (nearest second)

**5** 060°

**6 b** $(v^2 - w^2\sin^2\theta)^{\frac{1}{2}} + w\cos\theta$

**7** 26 m s$^{-1}$ bearing 023° (nearest degree)

**8 a** $10\sqrt{3}$ km h$^{-1}$    **b** 9.27 a.m. (nearest minute)

**9 b i** 571 m (nearest metre)

    **ii** 2.05 p.m. (nearest minute)

**10** $\left(-\frac{15\sqrt{3}}{2}\mathbf{i} - \frac{15}{2}\mathbf{j}\right)$ km h$^{-1}$

**11 a** 013° (nearest degree) **b** 131 s (nearest s)

**12 a** 337° (nearest degree) **b** 1306 (nearest minute)

   **c** 2.3 km (nearest 0.1 km)

**13 b** 1.27 h (3 s.f.)

**14 a** 11.5 (3 s.f.)

   **b** When $\theta$ is calculated using the sine rule the answer is ambiguous.

   **c** 1316

**15 a** $(3\mathbf{i} + 2\mathbf{j} + \mathbf{k})$ m

   **b i** $(-4\mathbf{i} + 6\mathbf{j} - 2\mathbf{k})$

    **ii** $(11\mathbf{i} - 10\mathbf{j} + 5\mathbf{k})$ m

**16 a** 4.82 m s$^{-1}$ (3 s.f.), 275° (nearest degree)

   **b** 240° (nearest degree)

**17 c** $\frac{1}{1 + 2\tan\theta}$

   **d** 20.6 m s$^{-1}$ (3 s.f.), 024.9° (3 s.f.)

**18 a** 265 km h$^{-1}$    **b** 344° (nearest degree)

   **c** $\frac{43}{53}$

**19 a** 61.9° (nearest 0.1°)    **b** 80 s    **c** $\frac{16}{3}$ m

**20 b** $q = 0.4$    **c** $p = \frac{8}{3}$

**21** $\frac{1}{2}mu^2(e^2\sin^2\alpha + \cos^2\alpha)$

**22** $\frac{1}{3}$

**23** $\left(-\frac{23}{9}\mathbf{i} - \frac{46}{9}\mathbf{j}\right)$ m s$^{-1}$

**24 a** $\frac{2}{3}$    **b** $\frac{1}{3}$

**27** $\frac{1}{3}$

**29 b** $45°$

**30 a** $A: (3\mathbf{i} + \mathbf{j})\,\text{m s}^{-1}$      $B: (2\mathbf{i} - 3\mathbf{j})\,\text{m s}^{-1}$
  **b** $4m\,\text{Ns}$
  **c** $37°$

**31 a** $A:$ parallel $1.5\,\text{m s}^{-1}$      perpendicular $2\,\text{m s}^{-1}$
       $B:$ parallel $0.5\,\text{m s}^{-1}$      perpendicular $1.2\,\text{m s}^{-1}$
  **b** $A: 2.1\,\text{m s}^{-1}$ (1 d.p.)    $B: 1.9\,\text{m s}^{-1}$ (1 d.p.)

**32 a** $\frac{3}{4}$                 **b** $11\,\text{m s}^{-1}$

**33 b** $\frac{\sqrt{3}}{4}(1 - e)U$ and $\frac{1}{2}U$    **c** $73.9°$ (1 d.p.)

**34 a** $\sqrt{2}mu$              **b** $d$

**35 c** $\frac{1}{32}mu^2$

**37 b** $\frac{4}{5}$               **e** $12:25$

**38 c** $\frac{40}{21}d$

**39 b** $\frac{2t}{1 + 3t^2}$        **c** $30°$

**40 a** $A: \dfrac{(1 - e)u}{2}\mathbf{i} + v\mathbf{j}$    $B: \dfrac{(1 + e)u}{2}\mathbf{j}$

**41** $4.5\,\text{m}$

**42 b** $2.4\,\text{m s}^{-1}$ (2 s.f.)

**46 b** $\frac{1}{2}m\left(U^2 - \frac{g}{2k}\ln\left(1 + \frac{kU^2}{g}\right)\right)$

**48 b** $\frac{m}{2k}\ln\left(\frac{8}{5}\right)$

**49 b** $20$            **c** $1600 - 1500\,e^{-\frac{t}{20}}$

**51 b** $18\ln\left(\frac{15}{30 - V}\right) + 21\ln\left(\frac{50}{35 + V}\right)$

**52 a** $16.1\,\text{s}$ (3 s.f.)      **b** $23.6\,\text{s}$ (3 s.f.)

**53 b** $\frac{g}{k^2}(1 - \ln 2)$

**54 c** $\frac{c}{k}\left(\ln 2 - \frac{1}{2}\right)$

**55 c** $\frac{\pi}{2}\sqrt{\left(\frac{k}{2g}\right)}$

## Exercise 4A

**1 a** $2e^{-2t}(\cos 2t + \sin 2t)$
  **b** $0.0901$ (e s.f.)
  **c** lightly damped

**2** $6e^{-2t} - 2e^{-6t}$

**3 a** $e^{-t}\left(\cos\sqrt{5}t + \frac{1}{\sqrt{5}}\sin\sqrt{5}t\right)$

  **b** $\frac{\pi}{\sqrt{5}}$ or $1.40^{\text{c}}$ (3 s.f.)

**4 b** $ut\,e^{-2\omega t}$

**5 b** $\frac{a}{2}e^{-\frac{1}{2}nt}\left(\cos\frac{n\sqrt{3}}{2}t + \frac{1}{\sqrt{3}}\sin\frac{n\sqrt{3}}{2}t\right)$      **c** $t = \frac{1}{2\omega}$

## Exercise 4B

**1** $-\frac{k}{8}\cos 3t + \frac{k}{15}\sin 3t + \frac{k}{8}\cos t$

**2 c** $\frac{V}{3\omega}(2 - 2e^{-3\omega t} - 3\omega t\,e^{-3\omega t})$

**3 b** $\frac{2U}{3k}e^{-3kt} - \frac{3U}{3k}e^{-2kt} + \frac{5U}{6k}$

**4 b** $\frac{a}{42}e^{-8nt} - \frac{a}{6}e^{-2nt} + \frac{a}{7}e^{-nt}$

**5 c** $e^{-\frac{1}{2}t}\left(-\frac{2}{5}\cos\frac{7}{2}t + \frac{48}{35}\sin\frac{7}{2}t\right) + \frac{2}{5}$
  **d** $5\pi - \frac{2}{5} + \frac{48}{35}e^{-\frac{\pi}{2}}$
  **e** $\frac{25}{4}\left(\frac{2}{5} - \frac{48}{35}e^{-\frac{\pi}{2}}\right)$

## Exercise 4C

**1 b** $\frac{5}{23}\sin 2t - \frac{\sqrt{2}}{23}\sin 5\sqrt{2}t$

**2 a** $e^{-kt}(A\cos\sqrt{(n^2 - k^2)}t + B\sin\sqrt{(n^2 - k^2)}t$

  **b** $\dfrac{2\pi}{\sqrt{(n^2 - k^2)}}$

**3 a** $l + \dfrac{g}{2k^2}$

  **d**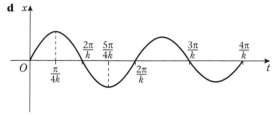

**4 b** $x = Ut - \dfrac{U}{n}\sin nt$

  **c** $\dfrac{2\pi}{n}$

  **d** $\dfrac{2U\pi}{n}$

**5 b** $\frac{6}{\sqrt{5}}e^{-t}\sin\sqrt{5}t$      **c** $0.514\,\text{s}$ (3 s.f.)

**6 b** $\frac{U}{k\sqrt{15}}e^{-\frac{kt}{2}}\sin k\sqrt{15}\,t$

## Exercise 5A

**1 a** $-mgl\cos\theta - 2mgl\cos\theta$
  **b** $\theta = 0$ stable
     $\theta = \pi$ unstable

**2** $V = -2mga\cot\theta + 4mga\,\text{cosec}\,\theta + k$
  $\frac{a}{\sqrt{3}}$ below pulley, stable

**3 b** $\theta = \frac{\pi}{6}$ unstable $\theta = \frac{5\pi}{6}$ stable

**5 b** $\theta = 0$ or $0.93^{\text{c}}$ (2 s.f.)
  **c** $\theta = 0$ unstable $\theta = 0.93^{\text{c}}$ stable

**6 b** stable

**7 b** $2 \leqslant k$          **c** stable

**8 b** $\theta = 0$ unstable    $\theta = \pi$ unstable    $\theta = \frac{\pi}{3}$ stable

**9 a** $0 < k < 2$
  **b** $k = \sqrt{2}$
  **c** unstable so will not return to horizontal position

**10** $\theta = \frac{\pi}{4}$ unstable    $\theta = \frac{7\pi}{12}$ stable    $\theta = -\frac{\pi}{12}$ stable

## Review Exercise 2

**1 a** $3\sin 2t - 6e^{-t}\sin t$    **c** $1.07\,\text{m}$ (3 s.f.)
  **d** $\pi$

**2 b** $\frac{\pi}{3}$

**3 b** $\frac{l}{4}(\sin 2\omega t + \cos 2\omega t - 1)$

  **c** $\frac{l}{4}(\sqrt{2} - 1)$

**4 b** $A = 0$    $B = \dfrac{U}{\omega}$    **c** $\dfrac{\pi}{4\omega}$

**5 c** $\frac{1}{2}ae^{-kt}(\cos kt + \sin kt)$

**7 d** $\dfrac{3u}{2}$                 **e** $\dfrac{\pi}{2n} + \dfrac{g}{un^2}$

**8**   **c**   $-Ut\,e^{-\frac{3}{2}t}$      **d**   $\dfrac{2u}{3e}$

**9**   **b**   $1.5L + \dfrac{Le^{-kt}}{2\sqrt{3}}\,(\sqrt{3}\cos k\sqrt{3}t + \sin k\sqrt{3}t)$

**10**   **b**   $\frac{4}{15}\cos 5t - \frac{1}{15}\cos 10t$

**11**   **b**   $e^{-kt}\left(\dfrac{V}{5k}\cos 3kt - \dfrac{4V}{15k}\sin 3kt\right) + Vt - \dfrac{V}{5k}$

**12**   **b**   $\frac{1}{2}ft^2 + \dfrac{f}{n^2}\cos nt - \dfrac{f}{n^2}$

     **c**   $\dfrac{2f\pi}{n}$

**13**   **b**   $0.3e^{-4t} - 0.6e^{-2t} + 0.3$

**14**   **b**   $C = D = a - \dfrac{g}{2k^2}$    **c**   $\dfrac{\pi}{k}$

**15**   **b**   $2e^{-2t}$      **c**

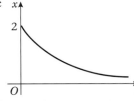

     **d**   Model not realistic as $\dot{x} = -4e^{-2t}$ and so $P$ is always moving ($\dot{x}$ is never zero).

**16**   **c**   $e^{-\omega t}\left(\dfrac{3u}{10\omega}\sin 2\omega t - \dfrac{2u}{5\omega}\cos 2\omega t\right) + \dfrac{2u}{5\omega}$

**17**   **b**   $26.6°$      **c**   stable

**18**   **b**   $0$ and $\pm\dfrac{\pi}{3}$

**19**   **b**   $\pm\cos^{-1}\left(\frac{3}{4}\right)$      **c**   unstable

**20**   **b**   $\dfrac{\pi}{6}$

**21**   **b**   $75.5°$

**23**   **b**   $\dfrac{\pi}{3}, 0, \pi$

     **c**   $\theta = 0$ unstable   $\theta = \pi$ unstable   $\theta = \dfrac{\pi}{3}$ stable

**24**   **c**   stable

**25**   **a**   $s - (L + 2L\sin\theta)$    **c**   $0.322^c$ (3 s.f.)

     **d**   unstable

**26**   **b**   $0$ or $\dfrac{\pi}{3}$

     **c**   $\theta = 0$ stable      $\theta = \dfrac{\pi}{3}$ unstable

**27**   **a**   Wire is smooth, so the reaction from the wire on the ring is vertical. If the ring is in equilibrium the tension in the string must be vertical as the third force on the ring is its weight.

     **c**   $0, \pi, \dfrac{2\pi}{3}$

     **d**   $\theta = 0, \quad \pi$ unstable,   $\theta = \dfrac{2\pi}{3}$ stable

**28**   **b**   $0, 1.32^c, \pi$

**29**   **b**   $0.841^c$      **c**   stable

**30**   **b**   $0.635^c$ or $-1.003^c$ (3 d.p.)

     **c**   $0.635^c$ unstable $- 1.003^c$ stable

**31**   **b**   $0.723^c$ (3 s.f.)      **c**   stable

**32**   **c**   stable

## Examination style paper

**1**   $v = \dfrac{u\sqrt{19}}{5}$

**2**   **b**   $\sqrt{2}\,\pi$      **c**   $0.881\,\text{m}$

**3**   **a**   $086°$      **b**   $0.31\,\text{km}$

**4**   $36.7\,\text{m}$

**5**   **a**   $\dfrac{\sqrt{10}u}{2}$      **b**   From $072°$

**6**   **a**   $5\mathbf{i} + 5\mathbf{j}\,\text{m s}^{-1}$      **b**   $2m(\mathbf{i} + \mathbf{j})\,\text{N s}$

     **c**   $1$

**7**   **b**   $\theta = \dfrac{\pi}{2}$ or $\theta = \sin^{-1}\frac{2}{3}$

     **c**   $\theta = \dfrac{\pi}{2}$ unstable, $\theta = \sin^{-1}\frac{2}{3}$ stable

# Index